MW01285069

This Is for Everyone

Also by Tim Berners-Lee

Weaving the Web:
The Original Design and Ultimate Destiny of the World Wide Web

This Is for Everyone

The Unfinished Story of the World Wide Web

Tim Berners-Lee

with Stephen Witt

Farrar, Straus and Giroux • New York

Farrar, Straus and Giroux
120 Broadway, New York 10271

EU Representative: Macmillan Publishers Ireland Ltd, 1st Floor,
The Liffey Trust Centre, 117–126 Sheriff Street Upper, Dublin 1, DO1 YC43

Printed in the United States of America
Published simultaneously in the United States by Farrar, Straus and Giroux
and in Great Britain by Macmillan
First American edition, 2025

Graphs on pages 86 and 100 and diagram on page 279 by ML Design Ltd.

Library of Congress Control Number: 2025940158
ISBN: 978-0-374-61246-7

10 9 8 7 6 5 4 3 2 1

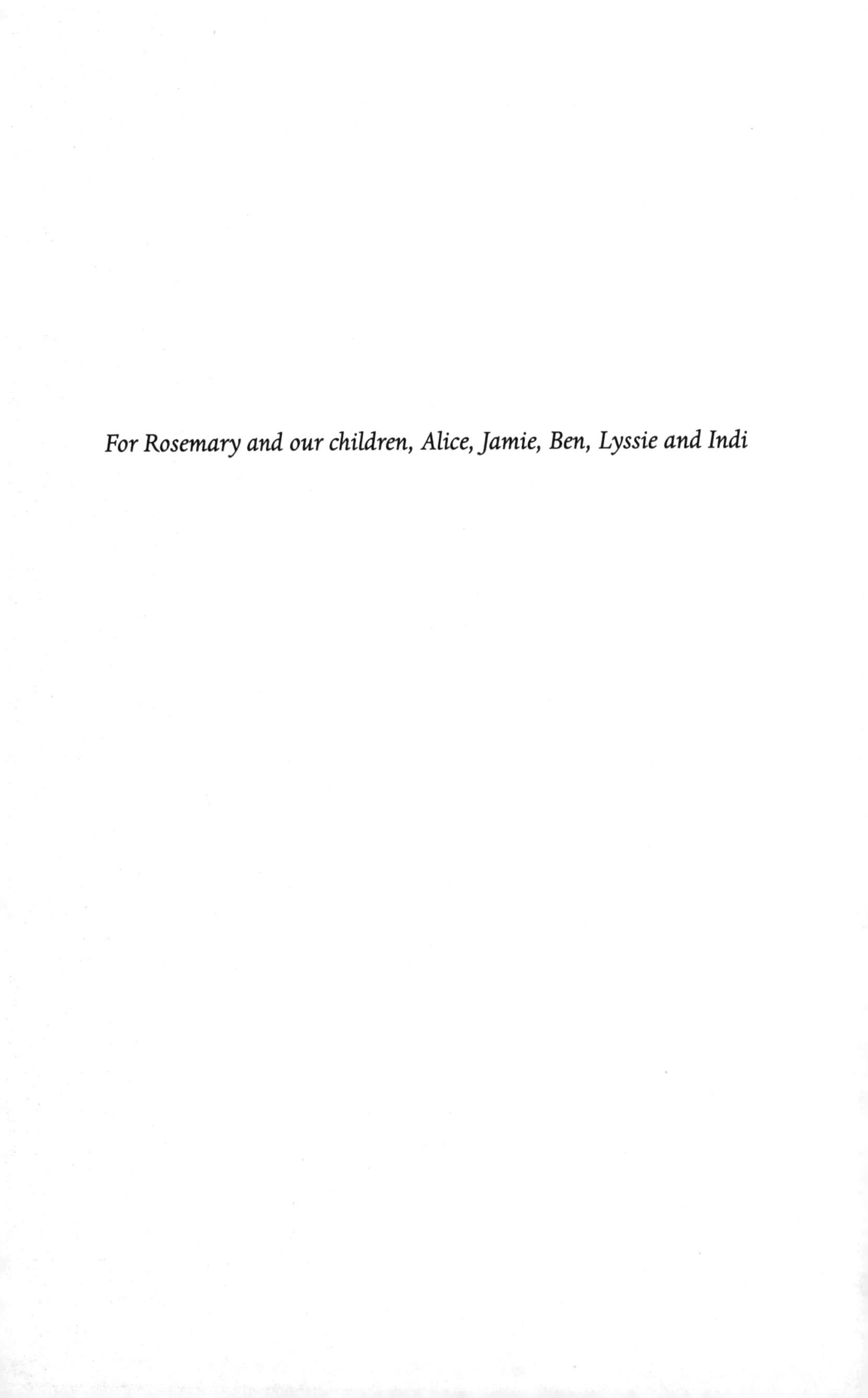

For Rosemary and our children, Alice, Jamie, Ben, Lyssie and Indi

Contents

CONTENTS

Prologue

I was thirty-four years old when I first presented the idea for the World Wide Web. At the time, I was working in Switzerland as a programmer at a particle accelerator. No one was asking for the web, and almost no one expected anything to come from it. I had never been to Silicon Valley, had no connection to venture capital, and was far from computer science research centres like Stanford University and the Massachusetts Institute of Technology (MIT). I had no track record as an inventor, held no patents, had never started a business, had never managed a team of people, and had published only a couple of research papers.

My job was at CERN (originally the European Council for Nuclear Research, now the European Organization for Nuclear Research), the international high-energy physics lab in Geneva, Switzerland, the largest and most complex laboratory of its kind ever built. CERN's mission was to discover the ultimate nature of matter – what is it? Where does it come from? To do this, CERN accelerated beams of subatomic particles on a gigantic racetrack, faster and faster, then smashed them into each other to see what happened. I helped write the code that allowed all the many computers and devices there to communicate with each other.

Working at CERN was pretty good fun. There were projects and people from all over, and lots of different sorts of computers, from mainframes in the computer centre, to workstations in the control centres, to microprocessors in the tunnel around the accelerator, which was located in a 27-kilometre-long circle, 300 metres underground. To keep track of all these complex systems, I needed – in fact, *they* needed – a powerful information system which would bridge those projects, those different types of computer systems, those teams, those people, and those ideas. (Did they know they needed it? Nope.)

I spent a long time thinking about how to do this. Then, around 1988, I was struck by the idea of combining two pre-existing computer technologies into a single platform. The first technology was the internet, which is a protocol for connecting computers; you've probably heard of it. The second technology was hypertext, which takes an ordinary document – like a technical manual, or a diary entry – and brings it to life by adding 'links'. My idea was that these hypertext links could provide a simple way for users to navigate the internet.

This kind of decentralized structure could have a network effect on creativity: new trends, possibilities, and products would emerge as millions of people started linking, sharing, and following across the planet. If you could put anything on it, then, after a while, it would have everything on it. Because my idea could connect so many people, systems and countries, I called it the World Wide

Web. But when I described this vision to people, most of them seemed to regard me as a little eccentric.

CERN's assignment was to discover the origins of matter, *not* sponsor experimental networking technology. Still, I relentlessly petitioned my bosses at CERN to fund the World Wide Web. I took every opportunity to discuss it, pitching it in meetings, sketching it out on a whiteboard for anyone who showed the slightest interest, or even finding ways to raise it in informal settings. My friend Meryl Dalitz recalls me drawing the concept of the web in the snow with a ski pole during what was meant to be a peaceful day out on the slopes.

I don't think Meryl could see what I saw. What she could see was my passion. I still have that passion more than thirty years later. Actually, I think I have *more* passion now than I did then. The diagram I drew in the snow melted, but my idea – the World Wide Web – is now used by more than half the world. It's one of the most successful inventions of all time. Just as I'd hoped, it has enabled a flourishing of human creativity and self-expression, and I think we're only beginning. In fact, I believe that the only limit to what can be created on the web is our own human imaginations.

In this book, I want to tell you the story of how the web came to be. How I finally secured some time to work on it from CERN, and how I built the first web page, the first web browser, and the first web server, all on a single computer in a small room on the second floor of the Computing and Networking building. How I met my early collaborators online, and how that small, single server expanded into a small network of servers, then to a large network of servers, and then a really large network of servers, moving so fast that by the end of a decade we'd taken over the globe.

I want to tell you the story of how the web evolved. I left CERN in 1994 and relocated to MIT, where for the next twenty-eight years I led a consortium that oversaw the advancement of the web from a primitive collection of networking tools into the powerful suite of technologies that now power much of our online life. Today, the web is everywhere, on everything, powering most of the applications you use and delivering much of the content to your mobile phone. It delivers streams of media to your television and serves as the front end for a trillion dollars' worth of global transactions every day. This didn't just happen – it took an enormous amount of collaborative work for the web to evolve in this way. That work continues to this day, powering new technologies like videoconferencing, augmented reality and, of course, artificial intelligence (AI), whose impact is only just starting to be understood.

AI breakthroughs are coming so rapidly that it's difficult to keep pace. From a technical perspective, what's happening is that software designed to mimic the human brain is 'evolving' on giant computers, some big enough to fill a warehouse and some surprisingly small. These AIs need data to train, and much of that data (especially for systems like ChatGPT) is sourced from the web. The resulting systems are so powerful they seem like magic.

In the near future, our lives will be transformed by AI 'agents' which will interact with the web and take actions to achieve specific goals. You might ask an agent to book you a vacation, or file your taxes; you might use it to tutor your children, or order your groceries. The agent connects to the web to complete these tasks, and in doing so, it constantly improves along the way. As you'll see in this book, I've been envisioning these types of web agents since the mid-1990s – I just couldn't predict what form they would take! The AI wave is one of our biggest opportunities yet, and promises to deliver a great deal of value for humankind. But experience suggests we also have to be careful; this technology is so powerful that it threatens dystopian outcomes.

Sadly, we've already seen how things might go wrong. Over the past few decades, I've fought to keep the web transparent, open-source and freely accessible. Unfortunately, in recent years, along with all the creativity, empowerment, and collaboration that I love on the web, a small, but significant part of it – the addictive forms of social media – have multiplied into something misleading, toxic and habit-forming. That's pretty far from my vision. But because this small part of the web is *so* addictive, people spend a lot of time on it, and as a result most web traffic is now concentrated in a handful of large platforms which harvest your private data and share it with commercial brokers or even repressive governments. That's pretty far from my vision, too. Worse, authoritarian governments now use the web to spread disinformation and surveil their own citizens, and that's as far from my vision as can be. While we use and celebrate the good on the web, we also have to address the bad.

In the age of AI, these threats are more urgent than ever before. To ensure the web agents serve people – not corporate profits, not governments, not *themselves*, but *people* – it's critical that we develop systems today that put the human first. In the early days of the web, I put my design tools in the hands of individuals, not administrators or corporations primarily motivated by profit. This was one of the best decisions I ever made. Today, we have to build tools and systems that empower individuals once again.

Fortunately, there are ways to leapfrog the web to a place where it is much better for humanity. When users have greater control over all their data, they can better resist the forces that are degrading their experience, and they can seek out new tools that can improve their lives. In my own work, alongside dedicated researchers at MIT, I've developed a system called the Social Linked Data protocol, or SoLiD for short. It permits users to take control over all the data in their lives and put that data together to achieve new results. It returns the web to its roots, giving creators new collaborative tools, while passing power back to users. As AI agents develop, they might even use Solid to create trusted platforms for delivering remarkable new services. 'Trust' is the key word there – in the world I envision, your relationship to AI agents will carry the same level of privilege and confidentiality that you'd expect from communications with a lawyer or a doctor.

When I was asked to transmit a live message to hundreds of millions of viewers at the Opening Ceremony of the 2012 London Olympics, I wrote: 'This Is For Everyone'. Now, in the age of AI, I believe that to be true more than ever. We can restore the web as a tool for collaboration, creativity and compassion across cultural borders. We can fix the internet, and the next person to build web tools can push the boundaries of self-expression and deserve our trust. I don't know who that individual is, and I can't predict where they'll come from. What I do know is that that passion is out there, and that if we dedicate our minds to it, we can take the web back. It's not too late.

Tim Berners-Lee

CHAPTER 1

Early Days

I was born in 1955, the same year as Steve Jobs and Bill Gates. Our cohort would ride the wave of computing technology like no other before or since. My father and mother were both mathematicians and electronic engineers. They worked at Ferranti, a Manchester-based electrical engineering firm that built the UK's first commercial computer. They met around 1950, while working together on the Ferranti Mark 1, and its successor, the Mark 1*. (The latter weighed 10,000 pounds.) The team was based in the 'Tin Hut', a lean-to erected along the outside wall of the electronics factory. Dad helped design and program the machine, while Mum wrote binary code with a tape punch. Her output was a roll of paper, with holes representing ones, and the absence of holes representing zeros – she was so skilled that she could hold the tape up to the light and read the code. While George Harrison's mother was pregnant with him, she would listen to Indian classical music on the BBC. While my mother was pregnant with me, she would take me on calls to computing customers to help them with their code. I recall as a young boy wrapping myself and my siblings in unspooled rolls of that same five-hole paper tape.

I was named Timothy after my father's cousin, who'd died in the war. My parents owned a semi-detached house in the London neighbourhood of East Sheen, between Richmond Park and the Thames. They lived there from 1954 until they died in their nineties. The house was close to Sheen's high street, the Upper Richmond Road, and its real shops – the butcher, the baker, the fishmonger and the greengrocer. I was the eldest of four children, and my mother, Mary Lee, paused her career once I was born, redirecting her considerable intellectual energy into raising us. She was an extraordinary woman, who encouraged creativity and curiosity in her children.

Early days at Emanuel School.

Our household was a tremendous amount of fun. We were lightly supervised, following a parenting philosophy Mum termed 'watchful negligence'. Mum would organize treasure hunts for special occasions, such as when we went camping. She hosted the best children's parties, with fantastic themes – I remember one year, she attached astronaut footprints to the ceiling for a birthday party in space. She created her own calendar, a large circular dial with 365 days arranged along the perimeter, and each year made a new face

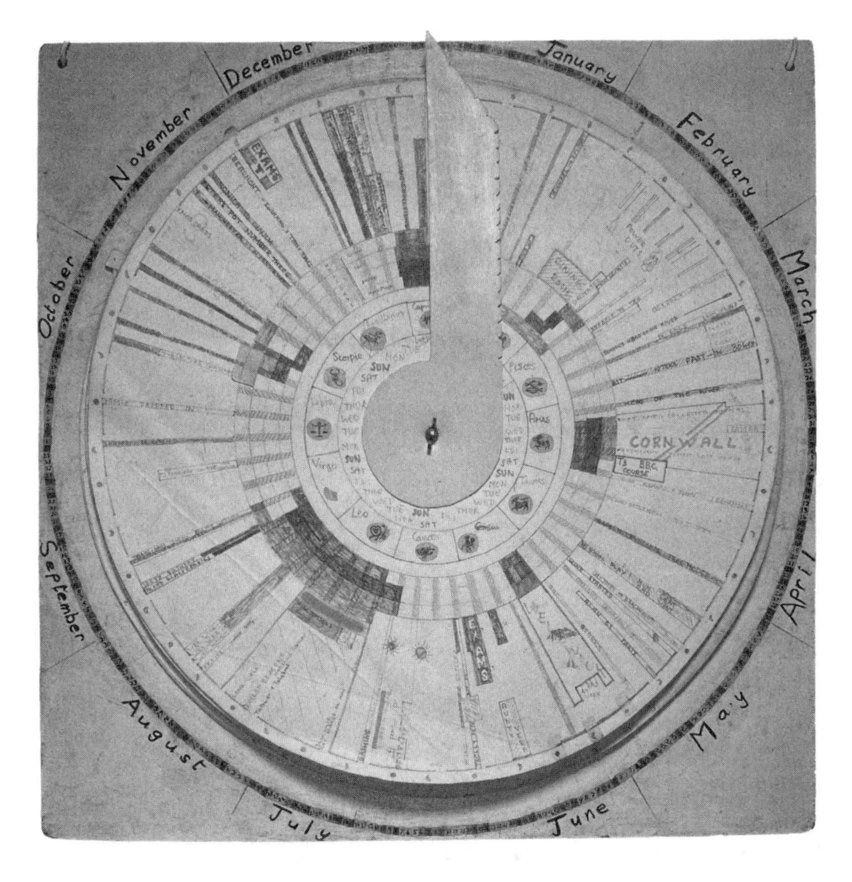

My mother's rotary wall calendar.

of card on which to record the year. She would add the things coming up in the year – especially school holidays – so we could see how far away they were. As the dial moved, she'd record notable events on the days, and when the dial had completed its rotation, you could see your whole year at a glance. It was a brilliant invention, superior to the typical wall calendar in every way. To this day, I keep a copy of Mum's circular calendar on the wall in my office, to remind myself of her, and to remember always to look for a different way of doing things.

Both my parents had done short 'war degrees' and both worked on electronics during the fighting. After the war, Mum went out to Australia, where she worked at the Mount Stromlo Observatory, near Canberra. She would head off into the bush in her time off, with a small tent made of ex-parachute material, and a toolbox on the back of her motorbike. She found she could fix most things on the bike with fence wire and pliers. That resilience from camping was a core strength she passed on: as a family, we would camp at National Trust sites or sometimes even in the gardens of friends. In fact, with three kids at the time, my parents once drove across France, camping all the way, to stay with a friend in Geneva. The knowledge that I can weather a storm in a small tent has been reassuring and grounding – it's the awareness that I don't actually need anything else.

Mum was also a crusader for equal treatment in the workplace. Around the time she joined Ferranti, the women at the company were furious to discover they were being paid less than men recruited at the same time. The women of Ferranti chose Mum to argue the case for equal pay. She secured pay rises for the female staff, and soon after that Ferranti agreed to pay everyone the same.

My father, Conway, was a mathematician and statistician who loved mathematical games. His mother was Helen Campbell Gray, a fashionable Canadian socialite who had moved to the UK as a nurse in the First World War, married a British soldier, and later wrote for *Vogue* magazine. When Dad proposed marriage to Mum, Helen initially disapproved, thinking she wasn't quite good enough for her son. Granny soon changed her mind, though – Mum had that effect on people.

We learned to enjoy mathematics wherever it cropped up, and learned that it cropped up everywhere. One aspect of Dad's job involved queuing theory, which is the study of the movement of people, objects or information through a line. Once, to make slides for a talk, he used me and my siblings to demonstrate queuing theory concepts: each of us stood with a few balls – representing tasks to be processed – kicking them from one to another. Queuing theory helped his audience understand how long each kid would wait for a turn to kick, and where a ball would be at any given time. My father also had a keen understanding of the limitations of computers and would often discuss the things that people could do easily, but machines not so well.

Dad was brilliant, but he could be a little absent-minded. One time, he took me to pick up his shirts at the dry-cleaner's. He got the shirts, but left me behind in my pushchair. Another time, he parked our car by an embankment on the Thames. He returned to find the tide washing over it. When he was travelling back to London from Manchester with colleagues one day, he could not find the return half of his train ticket at the barrier at the station. His colleagues assured the ticket inspector that he had just lost it and was always forgetting things. So he got home, and then my mother asked, 'Conway, where's the car?'

We four children – me, Peter, Mike and Helen, named after Granny – were close companions. Mum had a simple system for household chores, with each task performed by the youngest child capable of it. (This clever trick had the younger children begging to do housework.) We were a tight-knit family that did everything together, and we spent a lot of time outdoors. When we went to visit friends in the countryside, we'd bring our tents along, and the six of us would camp out on the lawn.

I attended Sheen Mount, the neighbourhood state school, clad in shorts, jacket and tie. My closest friends were Nick Barton and Christopher Butler. After school, we'd race to nearby Richmond Park, which comprised 2,500 acres of open space in London, populated with forests of old-growth trees and wandering herds of deer. Other boys followed football and listened to the Beatles, but I spent most of my time in the park. (I missed out on pop culture more or less entirely – when I met my wife, Rosemary, she was astonished to learn that I'd never heard of Bruce Springsteen.)

What culture I did absorb came through my parents, who were fans of theatre and classical music. I fondly remember our household reading of *The Importance of Being Earnest*, with Dad playing the role of the redoubtable Lady Bracknell. ('A handbaaaag?!') The focal point of the house was not the television, but instead a big old bookcase with glass doors, which held the *Encyclopaedia Britannica*, atlases, Dad's maths books and the complete works of William

Shakespeare. Also on the shelves, as would later prove relevant, was a fussy Victorian manual of practical household tasks, entitled *Enquire Within Upon Everything*.

•

At the time, the set of people at work on computing was small and it's little surprise my parents got to know Alan Turing, the godfather of computer science. During the war, Turing had been the UK's greatest codebreaker. He was a lean and handsome polymath who drafted a conceptual outline of the computer – known as 'the Turing machine' – several years before the hardware was available to build one. His team then put these ideas into practice at Bletchley Park, a stately English country house north-west of London, building the electromechanical contraption known as Colossus that cracked the German Enigma cipher. His work was kept confidential for decades afterwards, but historians now credit Turing and his team with shortening the war by several years.

While at Bletchley Park, Turing proposed marriage to Joan Clarke, one of his fellow mathematicians. He then broke off the engagement and admitted to Clarke that he was gay. He pursued clandestine relationships with men, although homosexual acts were illegal in the UK at the time. After the war, Turing worked with the Ferranti Mark 1, trying to teach it to play a passable game of chess. (He never succeeded. The punch-tape computer was too primitive, and researchers were in the very early stages of understanding which problems were easy for a computer, and which were hard. Chess was hard!) In this way, Mum and Dad came to know Turing personally.

In a 1950 paper entitled 'Computer Machinery and Intelligence', Turing published a seminal thought experiment he termed the 'Imitation Game'. The game imagined a computer capable of holding a conversation that was indistinguishable from one with a real person. Any computer program that could do so was said to have passed the 'Turing Test'. It would be many years before machines came close to passing the Turing Test – and when they did, the web played a critical role in making it happen. But I'm getting ahead of myself.

The early days of computing in the UK were portrayed in *The Imitation Game*, the 2014 Alan Turing biopic. When the film appeared, my parents, then in their nineties, subjected it to a rather withering fact-check, complaining both about the whitewashing of Turing's sexuality and the inaccurate representation of certain beloved wartime computing machines.

In 1952, Turing was convicted of 'gross indecency' after admitting to a relationship with another man – he finally received a posthumous royal pardon in 2013. He died in 1954, most likely by suicide. (Police found a half-eaten apple dosed with cyanide next to his bed.) I was born the following year. We never met, but his influence on my parents was profound. To be born in this time, in this place, with these unique parents, in this extraordinary family, was a massive privilege – though I didn't understand how special it was till later.

•

At eleven I enrolled at the Emanuel School, a secondary school located a short train ride from my house. Academic standards at the school were excellent, but it also had an optional programme of official conversion into the Anglican Christian faith, which I opted for. (Why? More to please the system than anything.) As part of that I was subjected to some rather intense fire-and-brimstone sermonizing. My mother was brought up as a Christian Scientist and converted to the Church of England when she met my father. She was quite devout, but she was an actual scientist as well: she thought every church should have the words 'Everything Within Should Be Taken Metaphorically' carved above the entryway. She was liberal in that she thought people should be allowed to do what they chose in private, so long as it did not hurt anyone else.

I tended to be a little more sceptical; I don't think I even believed in Father Christmas as a young child. As Richard Dawkins points out, religions tend to get the adolescent to commit to their faith just a moment before they have the level of rational thought to be able to see through the whole thing. So it was for me – my confirmation at Emanuel more or less coincided with my complete rejection of the whole faith-based edifice. The hellfire and damnation sermons of the confirmation classes may have been a last straw. Instead of the Bible, I became an avid reader of science fiction. I loved Heinlein, and Isaac Asimov's *Foundation* series. Stanley Kubrick's *2001* debuted just around the time my faith ran out. I saw it in cinemas, and have seen it many more times since.

Emanuel was a 'direct grant' school in British parlance, meaning it was run as a private independent school but was paid for by the government. The school had a distinguished legacy, having operated continuously since 1594. Emanuel was, alas, a boys-only institution, which must have delayed my social development immeasurably. Both my brothers went there too.

Although there was always food on the table, my parents were not wealthy. We never stayed in hotels, and buying school uniforms at Harrods was a considerable expense. I was lucky that when I got into Emanuel the government paid for my tuition. I was fortunate, too, that if I got two A levels, the borough would pay for any university, including Oxford. My parents made a contribution, but I basically got a free education, and I grew up imagining that the world would become more supportive of our kids, rather than less. Unfortunately, that hasn't proved to be the case: Emanuel, like many other direct grant schools, is now an independent private institution.

Primary school had been relatively easy for me, but as I prepared to move on to Emanuel, my mother cautioned me that things would be a lot harder at the next level, and that people would be a lot smarter. (When I went to Oxford, she would say the same. She was right.) My favourite subject at Emanuel was mathematics, taught by Frank Grundy, a terrific teacher who also happened to be an expert bridge player. Frank turned a blind eye to his students playing cards in the back of the class, though when the game was over he would deliver stern critiques of the play of the hand. Under his supervision, I studied for two of my A levels and also did maths for my Oxford entrance exam. Peter Jones, another maths teacher, took extra sessions in his lunch break to teach us vector calculus. I recall asking Peter why we were studying stuff that wouldn't even appear on any A-level exam. 'It's just so beautiful,' he said.

The chemistry teacher, 'Daffy' Purnell, was wonderfully encouraging and fun. The choice of A levels was typically between physics and either maths and chemistry, or pure maths and applied maths. The school tweaked the timetable so three of us could do all four. Daffy said he knew I was a mathematician at heart, but was welcome in any chemistry lesson. I think I was most nervous about the chemistry practical exam, where you were given a mystery substance and asked to determine what it was. On the day, after a bit of experimenting, I thought maybe it was sodium sulphite. Not really knowing much about related chemicals, I asked for two long rows of empty test tubes and filled the tubes in one row with chemical X and the others with sodium sulphite. Then I dropped every single liquid reagent I could find in the lab into the tubes to check they got the same result. To my chagrin, one of the tubes reacted differently! With no time left to find a plan B, I just confessed to failure on the exam sheet – but the examiners apparently liked the logic and I got an 'A'. There has to be a lesson there somewhere.

In tandem, I was receiving an education at home in computer science. One of Dad's jobs was writing speeches for Ferranti's CEO, Basil de Ferranti. ('Don't sniff at the sonatas of archdukes,' he used to say. 'You never know who wrote them.') He developed the skill of describing computers using simple analogies, like explaining binary mathematics with coins of a penny, halfpenny and farthing. One of my dad's favourite demonstrations was a cascading system of water jets he'd arrange, to simulate the way electricity flowed through a computer's circuits. By positioning the jets in the right way, you could create a 'logic gate' which varied its output according to simple rules. By chaining these logic gates together, you could start to do rudimentary calculations, like one plus zero or one plus one. And by chaining those primitive parts together, you could make a computer – a so-called 'universal' Turing machine, capable of all tasks accomplishable by computers.

Computers were exciting, but even then, I knew they could not do everything. I understood from talking to my father that they could record things in tables, but they could not typically remember random associations – passing connections, like when you smell a cup of coffee and are reminded of a trip to Ethiopia three years earlier. Although I was still young, I began to envision a way to design a program which could store these random associations – these arbitrary links. In fact, even as an adolescent, I sometimes wondered if perhaps it was these links, and not the objects they linked to, that were the important thing.

•

While attending Emanuel I decided I would build my own computer. What choice did I have? Personal computers like the Apple II were still years away, and even if they were available, I couldn't have afforded one. Besides, homebrew computing gives you a feel for the system from the CPU to the backplane. This was a project that would take me years to complete, and it helped lay the foundations for my journey. In fact, as I later learned, building a computer from scratch was a rite of passage for many engineers in my generation – Bill Gates and Steve Wozniak did it too.

I already had some experience with electrical engineering. With my friends Nick and Christopher, I had manufactured an electromagnet by softening an iron nail in the fireplace, then wrapping the nail in coils of copper wire. Attaching one end of this electromagnet to a battery, and the other end to a mousetrap, I had created a sort of makeshift gun that shot wooden missiles under remote control. Using electromagnets, I had also tried to build some elementary electrical relays, or switches, but there was a limit to what you could practically build with those.

Then the transistor arrived. Ah, the transistor! That miracle of twentieth-century engineering that conquered the world. The transistor – at the time, a quarter-inch-sized component with inch-long wires coming out of it – is a device that can alternate between on and off states to control the flow of current. In essence it is a simple, low-cost electrical switch. With just a few transistors, you can create logic gates, similar to the ones my dad had demonstrated with water, or like the ones we had made using relays. When I was fourteen, the Apollo engineers used these transistors to guide Neil Armstrong and Buzz Aldrin to the surface of the moon. Even now, transistors form the basis of all computing – they have just got smaller and smaller and smaller. Today's transistors are thinner than a strand of human DNA.

Thanks to the manufacturing boom in Asia, by the early 1970s transistors had plummeted in price. At a surplus store on Tottenham Court Road in London, I could buy a bag of 100 factory-reject transistors for just a couple of quid. Those transistors came in all shapes and sizes, and half of them were broken – to get a working batch, you had to go through and grade them, which was a painstaking task. But once you were done, you had the basic electronic toolkit to build pretty much anything.

The first thing I built was a switch for my model railway. Then I built an intercom that linked the upper and lower floors of the family house. ('He was very useful as the engineer around the house,' Mum would later say.) I bought a 'breadboard', a simple physical platform for building circuits, and started chaining together logic gates made from my cast-off transistors. You could make a circuit on the breadboard in minutes, and if it worked, you could solder it up on a printed circuit card to make it permanent. I made a train whistle circuit and some automation for the model trains.

The trains had been my pride and joy until electronics ended up taking over. Not surprising, as Emanuel was stuck in a wedge-shaped bit of land between two railway cuttings for the Brighton and Bournemouth lines as they forked away south from teeming Clapham Junction. In the majestic days of steam, you could hardly go to Emanuel without getting hooked on trains. You would hear the rhythmic clanking of the wheels on the rails and the high-pitched whistle echoing through the town. The engine itself, a magnificent piece of machinery, would hiss and puff as it powered its way down the track, trailing thick, white clouds of steam and coal smoke. Alas, in the late 1960s, the steam era came to an end, except for a handful of heritage lines. I'm glad I got to see it.

So from trains to electronics, from electronics to computers. At the time, though, computer science was not offered either at A level or as an undergraduate course at Oxford. Instead, I left school with A levels in maths (pure and applied), physics and chemistry, with O levels including Latin. Like many of the Emanuel staff, my maths teacher Frank Grundy was an Oxford graduate and, in retrospect, I see that he had been training me as potential Oxford material since the day I'd arrived. Still, I thought it prudent to at least consider

Cambridge, and one day I mentioned this to him. He blinked, then, very slowly, as if speaking from some vast distance, said, 'Yes, I suppose you could apply there.' Oxford it was, then.

•

Without computer science as an option, I had to pick something related. My parents had both read maths, and maths was my best subject at school. On the other hand, electronics was my passion at home and was much more practical. I reasoned that physics would occupy a middle place between the two. This turned out to be incorrect – physics is its own discipline, of course, with its own attitude and philosophy – but a happy choice in the end. I took Oxford's maths entrance exam, planning to read physics when I got there.

Oxford University is composed of a number of smaller colleges, and students work with dedicated tutors within those colleges for three years, as they pursue a degree. Parallel to this, the university departments provide lectures and labs. I applied to Queen's College as my prospective place of study. John Moffat, a renowned professor of nuclear physics, was the tutor there.

Having received an acceptable grade in the entrance exam, I was invited to an interview. On the day of the interview, I woke early to the sound of the bells in the Queen's clock tower and dressed myself in clothes I hoped would fit in at Oxford – a green corduroy jacket, a yellow shirt, brown trousers and a Marks and Spencer tie. (Do recall this was 1973.) I approached Moffat's office and found, to my slightly cringeful relief, that he was wearing exactly the same thing. I wasn't an Oxford academic yet, but I could play one on TV.

The interview went well. I got in with an 'exhibition' – a little scholarship. Over the years, Moffat would become a great mentor to me. We began with basic courses in errors, and maths for scientists. John Moffat taught us all different aspects of physics, like thermodynamics, quantum mechanics, atomic physics and so on, but not always in the same order as the university held the lectures. He didn't rely on the lectures – we could go to them if we wanted to, sure, but every week he would give each one of us some reading from the textbook, and some exercises to do for next week. I would take my problems away and work on them, and the next week he would go over my solutions.

John said I always had my own way of doing things: I would use my own unconventional notation in place of the standard variables, and I sometimes approached problems from an oblique angle. Moffat, to his eternal credit, would always just go with my flow and attempt to see things from my point of view. Over one problem set, Moffat spent quite a long time lost in thought, poring over my strange notation. Eventually, he raised his eyes and informed me I'd actually got the problem right – I'd merely dropped a minus sign just here, on the fourth page along the way. That ability to look at the whole thing from the point of view and vocabulary of another person was an incredible gift he had – and a hard one to emulate.

I loved Oxford, even if Queen's College, at that time, had not quite yet joined the twentieth century. The place was still entirely male, for one thing; Queen's didn't accept female students until 1979. We were required to wear gowns for dinner each evening, seating ourselves along long mahogany banquet tables with servers delivering our meals. At the front, on a raised dais, sat the faculty and post-docs in their robes. (If you've seen any of the Harry Potter films, you'll get the idea – the Great Hall at Hogwarts was shot at Oxford, though at Christ Church not Queen's.)

After dinner, the students would retreat to the excellent beer cellar, at the bottom of a staircase on Front Quad. Alcohol was a currency of Oxford life, though not the binge culture which I gather more recent generations have brought. One night sprawled on the lawn after overdoing drinks in the Provost's lodgings was enough to be a lesson for a long time.

The colleges don't have their own laboratory facilities, nor the large lecture theatres of the Physics Department, so we did our practical experiments and attended lectures there with students from the other Oxford colleges. From the front of the lecture theatre, the gender imbalance was pretty appalling. But in one of those early lectures the very first week – I think it may have been a lecture on errors – I met my first girlfriend, a brilliant undergraduate pursuing the same degree. We became increasingly inseparable, and soon we were a couple.

At the beginning of my second year, I moved into a college room overlooking Queen's Lane, above a coffee shop which billed itself as the oldest in Europe. I would awake to the smell of stale coffee, jolted from my slumber by the rumble of double-decker buses shifting into second gear.

Architecturally speaking, Oxford was a place wholly dedicated to the glory of God. Portraits of great theologians and the late archbishops of the Church of England were hung in the dining hall, and the colleges offered special scholarships for the student organist and choir singers. Chapel was held in a splendid domed hall, with light streaming in through stained-glass windows decorated with beloved saints. But my rational mind was pushing me in the opposite direction.

I began, in a friendly way, to debate with the chaplain. Much of the wisdom of the Church was still useful, and church services still had a lot of connections and value for me. I did not feel awkward going to the chaplain's coffee morning if taken by a friend. The chaplain at Queen's was, in fact, a good resource; I used him once as a data point on culture. 'Pride,' I asked, as we crossed in the quad. 'Definition, and is it a good thing?'

'Self-respect gone to seed – and generally not,' he replied. Noted. Little did I know how, decades later and an ocean away, I would be immersed in a culture which treated pride as a very important thing.

The Church itself seemed self-interested, perhaps even a little predatory, and I soon declared myself an atheist. However, a form of religion where you could have your wise chaplains, your celebratory services for joy or sorrow, your music and your fellowship, but without having to believe six impossible things before breakfast – that would have been handy. I got *my* moral instruction from Moffat, who even today remains one of my role models. Some undergrads had a moral tutor, for well-being and behaviour, and a separate academic tutor, but Moffat was both for me.

With the other physics students from Queen's, Lady Margaret Hall, Keble and Catz (St Catherine's), we formed a tight cohort. We would take punts out on the Cherwell, the smaller pretty river which flows through Oxford to meet the Thames. These punts are flat-bottom wooden boats, propelled by a long pole from the rear; moving one is a practical lesson in physics. You grip the slats of the boat firmly with your feet, hoist the huge pole to a standing vertical, drop it to the river bottom, and push to get the boat going in just the right direction. Your party then progresses gently under the willow branches, clad in wide straw hats, holding guitars, textbooks, picnics and lecture notes, trailing a cooling bottle of wine on a short rope.

•

The eight-week terms were full, so my electronics hobby mostly happened in the breaks back in London. By this time, I'd outgrown the factory-reject transistors and moved to general-purpose integrated circuits. The leading vendor at that time was Texas Instruments, which sold logic controllers that slotted directly into the breadboard. The 7400-series TTL chips were available in more and more combinations of functions, and powered a remarkable array of early consumer electronics. Each chip was about the size of a paper clip and cost only a couple of pounds. The wiring of the 7400s was hidden in ceramic casing, so arranging the right controllers in the right combinations along the breadboard meant consulting a series of circuit diagrams, and imagining, in the mind's eye, the flow of electricity through the transistor gates within.

A computer itself was going to require a serious, perhaps impossible, number of 7400-series chips, so a more humble but still satisfying goal was to make a computer *terminal*. The terminal is the bit that talks to the person – in the physics lab the terminals we used to talk to the computers were 'teletypes', which typed their output at ten characters per second on long rolls of paper. Teletypes were telegraph machines adapted for computers, back when telex was the fastest way of getting a message across the world. At this time, though, paperless terminals, with a keyboard and screen, called Visual Display Units were becoming more popular.

With the help of one of my classmates, Peter Gilyard-Beer, I decided that I would build my monitor out of an old television. I tracked down a local television repairman and asked if by chance he had any sets with broken reception but working screens. He sighed, yes, he certainly had, and directed me to a tower of such sets in the back of his shop. He sold me one for next to nothing. In fact, I got two, just in case.

In those days, televisions worked via a scanning 'electron gun' which shot beams through the vacuum tube (or cathode-ray tube) of the set onto the screen, moving from the top left corner and snaking down to the bottom right, refreshing fifty times per second. With Pete's help, I found the spot on the circuitry of the TV where I could attach a wire to inject my own video signal in place of the broadcast signal. I was doing all this in my room in the attic, but I left the second TV on the hall table downstairs and drove it from the same signal, so my parents could see my progress. They could appreciate, with the electronics they had each done in the war, that moving from horizontal to vertical lines on the screen was an achievement. From there, I moved to a grid of 16 lines, each with 64-character places.

Once I had the basic screen mapped, the final step was to wire up a keyboard. Here I got a little lucky. I'd taken a summer job at a timber yard, and, while disposing of a large barrel of wood shavings one day, I came across a discarded old adding machine in the bottom of a huge skip. The machine was ancient and consisted of

100 button keys arranged in rows of ten. Crucially, though, each key had a dedicated electronic switch. I painted over the buttons in white and layered the alphabet on top, using transfer letters. I then attached a small circuit board to each button, to convert its electromechanical output into binary codes that the breadboard logic controllers could read.

So I had a terminal. I brought it into the physics lab, and the engineer of the PDP-8 minicomputer allowed me to test it, on the condition that I built an 'optical isolator' (a gadget with LEDs and photocells and an air gap) so he could be sure my nasty terminal could not harm his precious computer. Fair!

It was neat in a way that throughout this whole process, just as I needed something which I could not build myself, like a transistor or, now, a memory chip that could hold each of the 1,024 characters in my new display, industry produced it at a cost I could afford. That's why I feel my cohort – not my generation of baby boomers but more specifically the kids born close to 1955 – had a magic-carpet ride through the wave of tech. A few years earlier, everything would have been much too expensive or impossible. A few years later, it was all taken for granted and old hat.

My terminal couldn't run code, but in my last year at Oxford, Motorola introduced its 6800 'microprocessor', which was a whole computer on a chip. Oh, this was a beautiful little thing: a forty-pin, 2-inch package in maroon-and-gold housing that gleamed in the light and contained the actual Turing machine – the thing which executes instructions stored in memory. Rather than put it on one single motherboard as many home computer users were doing, I had a 19-inch-wide rack which I could upgrade just by plugging in cards. The M6800 came with a primitive 'assembly' language for designing and executing simple programs. Towards graduation, I began to show off my device – a fully operational computer that I'd built out of literal rubbish.

•

At Oxford, one's grade for the entire three years is determined by 'finals', a series of gruelling sit-down in-person examinations, which take place in little over a week. As the date of the first test approached, I set aside my microchips, eschewed the beer cellar and threw myself into full-time study. (My girlfriend and I were sharing a flat with our fellow students Janet and George, all frantically trying to ingest the same physics.) Clad in the formal academic gown with white shirts known as 'subfusc', we approached the Exam Schools on the High Street, which we had last seen two years ago, during the first-year exams. I sat down among a field of small, evenly spaced tables and opened the test paper.

After exams, we let our hair down. We broke onto the roof of the college along with other Queen's physics classmates. We sneaked across to the clock tower, decorating its hands with a pair of undergarments we'd found somewhere. We then inscribed the tower with Euler's formula ($e^{i\pi} + 1 = 0$), in an attempt to deflect suspicion onto the Mathematics Department. (It's important to get this stuff out of your system when you're young.) A few weeks later, I got a neat letter from Moffat with my results. 'Congratulations on a well-deserved first,' the letter read. It was as much his accomplishment as it was my own.

Following final exams I married my girlfriend Jane at the ripe old age of twenty-two. Nowadays we would have lived as partners before marriage, but things were different back then. With a first from Oxford, I could pursue a graduate degree in physics at pretty much any institution in the world. And yet, as graduation approached, I was coming to realize that academics was not my path. I loved Moffat, but I had no role models of people doing doctorates in physics and actually enjoying it. Perhaps this was a reflection of the era – particle physics was beginning to stall in the late 1970s – but the physicists I knew just didn't seem very happy. I'd had the time of my life at Oxford, though physics was never more than a detour for me. My destiny, whatever it was, had to involve computers.

If this story was happening today, I think what I might have done is go to Stanford University. Or if not that, at least been part of that scene of people who were building computers. But I didn't do that, simply because I didn't know that choice existed. In 1977, no recruiters from Silicon Valley came to career day at Oxford. In fact, back then, I'm not sure if I even knew Silicon Valley existed.

Instead, the recruiters at Oxford consisted of a few consulting companies and the three big telecommunications companies: GEC, ITT and Plessey. Wearing a boring grey suit with a tie, I did a round of interviews, but our decision was ultimately about location. GEC was in Coventry, which had been bombed out in the war, then rebuilt in concrete. (The interviewer said that Coventry was great because it was so easy to get away from it.) ITT was in Harlow, in Essex, a depressing 'new town' organized into commercial and residential sections, with 'green wedges' between them. Plessey was in Poole, a beautiful old fishing town on the south coast where the sea connects to a huge undeveloped natural harbour between the town and the Purbeck Hills. When Jane and I got off the train at Poole station, there were daffodils coming from the ground, and you could hear, already on the platform, the seagulls and the clinking of halyards against the masts of the moored dinghies. Sold!

To live in Poole, we had to work for Plessey. Near the Poole offices they were building telephone exchanges, and fancy fault-tolerant computers to run them. That is where Jane ended up. There were also some older buildings where the 'Private Communications Systems' division was building new things out of microprocessors. They showed me how a bar-code scanner, together with a cassette tape player and a modem, could let Sainsbury's supermarket switch from having a floor's worth of stock above the shop to on-demand ordering. That is where I ended up, working on new library automation systems.

Our lives in Poole were comfortable. We walked the coast paths often with our visiting friends Paul Rouse and his wife Lisa. Paul had been the other person at Oxford building his own computer, but I didn't meet him until just after finals. When we did meet, we became instant and long-term friends. We watched the sea crash into the cliffs at Winspit, and strolled along the limestone hills. On one walk in 1977 – a rainy walk along a windswept ridge by the sea – Paul, whose maths was much better than mine, took me through a recent paper by the cryptographers Rivest, Shamir and Adelman. The paper showed how you could use prime factors to create two huge numbers, one of which you write on the wall and tell everyone, and the other you keep very secret. This was 'public-key' cryptography, which took advantage of the fact that multiplying two primes is easy, but reverse-engineering the original primes from the product is impossibly hard – like trying to unscramble an egg.

That was a mind-blowing moment; it changed the world for ever, too. ATMs and remote logins and anything secure over the internet now depend on that RSA paper, as it was known. Years later, when we added the 's' in 'https' to make the web secure, that was all possible because of that paper. This encryption made secure online messages possible. I was astonished by the beauty of it.

Part of the beauty of it was that it gave power to a person as an individual. If they had existed back then, Paul and I could have each got out our laptops, made secret keys which we kept very safe, and exchanged the public keys, and then Paul would be able any time in the future to send me a message so no one else but me could possibly read it, and I would absolutely know that it came from Paul. No matter what email systems it went through, no matter what people tried to do with that message.

This technology gives you power as a sovereign individual – which is powerful, if you are a normal person in a rough world. Of course, it is also really useful if you are a spy. Once the US government noticed that, it classified the software as a munition, making it illegal to store it anywhere where it could be seen by a non-US person. This was the beginning of a long fight between the individual computer user and the centres of power. That battle is still going on, and we will return to it many times in this book.

•

The 1970s were economically challenging for the UK. The country faced high inflation, rising unemployment and economic recession. Strikes and industrial action were common, reflecting widespread discontent among workers. The country's traditional industries, such as manufacturing and shipbuilding, were in decline, leading to job losses and urban decay in some areas. But for Jane and me, after three rich and stimulating years at Oxford, it was a time of widening horizons, of growth. And in computing, the bandwagon Jane and I were jumping on, it was an exciting time too. At Plessey, the technology was primitive by modern standards, but the work was leading edge back then. When I started, in late 1977, my workstation had a paper-tape reader and punch. Disk storage at work took the form of 8-inch floppy disks, but if you wanted to store a program, you could also record the data as a sound onto cassette tape. Back at home, I upgraded my homebrew computer to match Plessey's Intel-based systems. With a little engineering, I connected a paper-tape reader to the system, allowing me to run simple commands written in 'Timpl' ('Tim's programming language'), my own variant of the very simple programming language Forth. In a feat more of carpentry than of engineering, I got the pockets to automatically fold the tape by tracing the shape of the professional tape reader at work and crafting it out of wood. Treasured things – I wish I had kept them!

Another engineer who joined around this time was Kevin Rogers. He and his wife Karalyn would head back to see family in deep Wales, along the westernmost bit of the Pembrokeshire coast, and sometimes Jane and I would tag along in the back of their Mini. The winding road along the Welsh coast was beautiful country, dotted with ancient villages and pubs. After a couple of years at Plessey, Kevin called and said he was looking for help. He had joined D.G. Nash, a three-person start-up focused on dot-matrix printers, located in an industrial estate in Dorset, a twenty-minute drive from the Plessey offices. I would soon join as well.

The dot-matrix printer was a device which was made entirely out of hardware. A bit like my homemade terminal, the printer made letter patterns on the paper by using a hardware table of letter shapes, but now it was crying out, like so many things at that time, to have the logic board ripped out and replaced by a microprocessor system. The D.G. Nash version of the printer had all kinds of fonts, and graphics. It would emulate a typesetting machine so that a printer, before setting a page of text on expensive bromide paper, could run it through the 'proof printer' and check for errors on cheaper stock. It would do network protocols too: if you didn't want to pay top dollar for a real IBM printer, well, it would pretend to be one.

Working at D.G. Nash was enjoyable. I could have stayed there, improving printer technology on the job and observing the larger computer universe from there. But then I was offered a one-time gig that would change the course of my life.

CHAPTER 2

CERN

The offer was to work at the European Council for Nuclear Research, or CERN in 1980. (It's now called the European Particle Physics Laboratory, but the acronym CERN has stuck.) Straddling the border between Switzerland and France, CERN ran a number of particle accelerators for the high-energy physics community. The facility was a mix of large utilitarian buildings and underground tunnels housing some of the world's most advanced scientific equipment, set against the serene backdrop of the Swiss Alps. CERN's mission was to study the subatomic building blocks of our universe known as particles – protons, neutrons, electrons. This was done by accelerating streams of these particles inside gigantic circular racetracks, then smashing them into one another.

The smallest racetrack at the accelerator complex was the Proton Synchrotron Booster (PSB), which injected proton streams into subsequent rings. CERN was running behind in rebuilding the control system for the PSB and had called in a team of British programmers to help. Previous accelerators had been controlled by complex electronics – oscilloscopes and detectors and sensors that looked like they'd been pulled from the laboratory of Dr Frankenstein. Part of our job was replacing these devices with software. As the new computer-based systems were being built, all of those lovely, glowing, electromechanical contraptions were being replaced by a single, boring ring of computer consoles. Such a pity, in a way.

CERN was a special place by any measure, but the country was not new to me. A few years earlier, I'd had a memorable stint at the European Semiconductor Equipment Company, just outside Zug, a pretty town on a pretty lake in Switzerland. The company manufactured a device called a wafer saw to cut silicon into chips, and they needed help with the microprocessors that controlled the tool, just as Nash had used microprocessors to control their printers. For an English kid who had never worked abroad, it had been pretty exciting.

And it had been fun working in Switzerland too. It was fun writing the code, and debugging it. It was fun staying with one of the company's engineers in a nearby village, even though it was a bit smelly due to the farm across the road. I had been surrounded by people speaking Swiss German, and much of the time is now a blur, but there were some indelible moments. In particular, I remember one night when we had had a long evening of visiting each other's homes, followed by restaurants, bars and dancing. We found ourselves on the upper floor of a chalet, with cows housed in stalls below us on the ground floor. We were drinking coffee, when the smiling grey-haired owner arrived with an industrial-sized container of his own best schnapps. Suddenly, the coffee became

Kaffee Fertig, the classic Swiss hot drink made of coffee, sugar and liqueur. It was a night of new and old cultures. I think it was then that I first fell in love with Switzerland. I'm still in love with it today.

•

I was delighted to be back on the continent at CERN, where Kevin, Jane and I were among the team sent to work on the centre's code. We had six-month contracts: July to December, 1980. Brian Carpenter, a blond Cambridge physicist who had received a PhD in computer science from Manchester, was the system manager. (Down the line, he would have a major role working with the internet and the structures around it.) It was a good gig – not your average office job. We started off working in the Terminal Room. At the far end were glass doors to the Machine Room, where the Norsk Data minicomputers actually sat. The term 'minicomputer' will seem like a misnomer to modern readers – each machine was the size of a refrigerator, with standard 19-inch-wide racks that ran from the (false) floor to the (false) ceiling. To the left of the Machine Room was a kind of mission control, where fancy, purpose-built screens with keyboards and trackballs allowed you to interact with any part of the particle accelerator system, such as vacuum pumps, accelerator magnets and, when all went well, the beam of particles itself.

Now, imagine what this place was like. Turn around in the Terminal Room from looking at the Machine Room, walk out of the door and you are in space – standing on a steel catwalk, a three-storey-high balcony, looking down across a grand experimental hall, the size of an aircraft hangar, with you in the top corner. Go along the catwalk and below to the right you will see equipment and concrete blocks, while through a window to the left you can see the CERN boundary road and behind it rows of Swiss vineyards. Follow the hall to the end, and you cross into a series of underground corridors that take you to the control equipment, arranged in a ring over the accelerator itself. The PSB was a 'starter mechanism', which meant it was a smaller particle racetrack about 50 metres across. It got the proton stream going, and, once it was up to speed, it fed it into the much larger collider beneath.

Come back along the catwalk, back past the Terminal Room on your right, and you are inside an office building, with corridors leading away at right angles. At the centre of these corridors there is a little coffee counter, and around the coffee place, in the intersection of the corridors, a number of small, standing-height tables. This little coffee place was, in fact, really important. For one thing, for us Brits, it was amazing to have real coffee, and real croissants. But also, when you have been parachuted into a running project, you need to be able to figure out all the parts of it on the fly.

Now in those days, different computing systems couldn't talk to one another. The problem was acute in academia, where scholars from competing computing paradigms segregated themselves into incompatible fiefdoms. And it was especially bad at CERN, where scientists from more than twenty countries were gathered in pursuit of a common goal, without a common language. One scientist might have critical information about how to run the accelerators stored in French in a private directory in the central Unix mainframe; another might have information on how to calibrate the sensors stored in English on an 8-inch IBM floppy disk in a locked metal cabinet; a third might have the data from the latest experiment on a printout in German beneath a coffee cup on his desk. It was a mess.

So, the only way to figure out how to run a particular device was to find the person who managed it and ask at coffee! He or she would explain how to use it, and also let you know who else you need to talk to, and then maybe that person would walk by, and they could pluck them out of the stream and introduce them. The informal vibe of that little coffee counter, with all these brilliant people collaborating across languages and technical disciplines – I wondered, was it possible to recreate this functionality with a read–write information system? I wanted a way of recording this information, *exactly as it came*, in random bursts, eventually building up a picture of the whole system, but with anyone, at any time, able to add new knowledge about anything.

So I wrote a program to run on the Norsk Data computer, on the standard 24-line, 80-character terminals, that allowed you to create notes about things. For each thing, you had to write notes about what type of thing it was: hardware, text file, report file, code file, paper document, concept or person. When you introduced a new thing, the only way to do it was to find a prior thing, and then say how the two things were connected – was a person, say, the author of the document, or the subject?

I called the program 'Enquire-within', short for *Enquire Within Upon Everything*, that book in my parents' bookcase. There were a few things I liked about it. It could easily switch between different languages, say French and English. Also, it would allow you to capture how any two things were related, without forcing you into a fixed structure. As often as not, information is presented on the page as either a tree or a table. It is organized into trunks and branches, or into columns. In my program, by contrast, you had a sprawling graph of links – something that looked a little like a web. That was in the spirit of those things my father and I had talked about: things that people could do easily but machines not so well.

When my time at CERN was over, I saved the Enquire program to an 8-inch floppy disk, then entrusted it to Brian Carpenter. I included with the disk a five-page written explanation of what it was, and what it did. It wasn't the last time I would see Brian Carpenter, but it was the last time I would see the disk. Our contracts were up and we returned to Poole. Sometime after, Jane and I decided to end our marriage and go our separate ways.

I sometimes wonder what my life might have looked like if I had not had the opportunity to go to CERN. I was fortunate to have the chance to work and live in Switzerland, out of my previous culture and comfort zone. I had a seed inside that was nurtured in that environment of cooperation and collaboration – a seed that sprouted into a tool that shook up the world. So, if you are given the opportunity to work abroad at thirty (or at any age really), my advice would be to take it.

•

Paul Rouse, my fellow Oxford computer-builder, was at that point in Geneva, working for CERN. He had applied for a two-year fellowship, and got to live in Switzerland and France, working on the latest systems with an international crowd of engineers, computer scientists and physicists. Attracted by a return to CERN, in 1983 I applied for the same two-year fellowship. In my application I made the mistake of saying I should maybe work on the forthcoming Large Hadron Collider, the entity CERN is best known for today. This sounded ambitious, but turned out to be a miscalculation: the LHC was at the digging and construction stage, so I was offered a job maintaining a database of concrete. I turned that down, then applied again the next year, being much more specific about the sort of stuff I would enjoy – microprocessors, systems communicating over networks, real-time stuff. In 1984, I got an offer I wanted to accept. I was a bit sad to leave the UK, but CERN gave me another chance to be around brilliant academics, and to work on challenging technical problems.

Great Britain was going through a rapid transformation at the time. Following the dreary economic experience of the 1970s, in 1979, the government began a series of radical economic reforms: lowering taxes, weakening labour laws, removing currency controls and gutting regulations. This was 'neoliberalism', and it rose in parallel in the UK and US before spreading to most of the rest of the world. At the time, the for-profit computing sector mostly consisted of hardware vendors. Software, by contrast, was considered freely reproducible code, often built collaboratively, with no expectation of profit. But starting in the late 1970s, entrepreneurs (most notably Bill Gates) began to develop software and evolve the industry. If you enforced copyright protections strictly enough, the profits generated from software could be enormous. By the mid-1980s, Gates's firm Microsoft, powered by its command-line MS-DOS product, was on its way to becoming one of the most valuable firms on Earth. (For younger readers: you used to interact with computers by typing in commands, one at a time, like a chatbot now.)

Much of my programming was done in Unix, an operating system that had been developed at Bell Labs in the 1960s. Unix programmers wrote modular, abstract tools to 'do one thing, and do it well', in the words of one designer. Unix also allowed the output of one tool to easily be routed as the input to another, enabling rapid and elegant multifunctionality. That was the design philosophy I always favoured, and it certainly influenced my later code.

But in the marketplace, Microsoft had rapidly outpaced Unix, introducing a schism in the computing community. Many of us considered Microsoft products inferior. Open-source, non-profit collaboration seemed to produce much better tools, and this mode of production didn't lock users into coercive relationships with monolithic Silicon Valley firms. We stuck to the old methods of software engineering, collaboratively pooling our resources, our time and our ideas with no expectation of remuneration. In fact, for much of my lifetime, programmers routinely built terrific software products for no money at all. They were talented people doing the work of ten normal programmers, who in modern-day Silicon Valley would command seven-figure salaries. Back then, they all worked for modest salaries; sometimes they worked for free.

•

I returned to CERN to find Enquire completely unused, and the floppy disk that contained it had been lost. (I heard that an intern had been given the disk by Brian Carpenter at one point, and liked the way the program had been written. If you find it, do let me know – I suspect it's now worth something.) In any event, programming accelerator control systems was no longer my assignment. Instead, I got to work in the Computing and Networks division to help with real-time data acquisition systems, now running on a new generation of Motorola-based hardware. This was a bracing engineering challenge, and one I took to, at least initially, with great enthusiasm.

Construction was underway on the Large Electron–Positron Collider. The LEP was a giant ring-shaped device, located in a circular, underground tunnel some 27 kilometres in circumference, straddling the Swiss–French border. A series of magnets accelerated the particle streams inside the ring to extraordinary speeds, then collided them in the middle of a detector in the hope of generating exotic, never-before-seen states of matter. The apparatus was so large that, standing in one of the tunnels, you could just make out the curvature of the walls as they stretched far into the distance.* (CERN is an amazing place – if you have not visited it, you should.)

I settled into a comfortable life in Geneva and nearby France. I rented a second-floor apartment, accessible only via a spiral staircase, in a block of flats in St Genis, just across the French border. Living in a French town was very different from Geneva city living. I frequently went windsurfing on nearby Lac Leman (or Lake Geneva, as anglophone readers may know it), and took up downhill skiing as well. (One thing I remember from this time is how difficult it is to get a windsurfer up a spiral staircase.) I joined the Geneva English Drama Society and performed 'Shipoopi' in a production

* The Large Hadron Collider would later be built in the same tunnel.

of *The Music Man*. My parents kept up correspondence by sending me letters written on reused computer paper, with my mother's neat handwriting on one side and printed lines of marked-up code on the back.

While living in Switzerland, I met Nancy Carlson, an American working as an analyst at the World Health Organization headquarters in Geneva. We moved in together and, a few years later, we married. Nancy worked with computers, too, and at home we shared a 286 Toshiba laptop. The personal computing revolution of the 1980s was well underway, and it was no longer necessary to build your own computer. I left my beloved homebrew computer behind in Poole; in the mad rush to leave for CERN, I ended up throwing it out. (Kevin, the photographer of the group, took some photos of it.) In my apartment in Switzerland, I was content with a PC clone.

The PC revolution put computers in the hands of ordinary people all over the world. In 1982, Steve Jobs appeared on the cover of *Time* magazine for the first time. These PCs had extremely limited functionality compared to the computers of today. They usually had single-colour monitors, often had no hard drives, and executed primitive programs from floppy disks. Connectivity, when it existed at all, was to Bulletin Board Services (BBS), made possible through slow-speed modems that could tie up your home telephone line for hours.

The BBS communities were small clusters of like-minded users, overseen by 'system operators', or SysOps. To connect to a BBS in the 1980s, you booted up your terminal software, typed a command to dial a specific number, then waited for the jagged, screeching tone of the modem to signal a successful connection. Greeted by a text-based welcome screen, you navigated through menus using keyboard commands, exploring message boards, downloading files, playing simple text games or chatting with other users.

The BBS systems were the first public online communities. They were fun, and beloved by users, but as they couldn't talk to one another, each one essentially was an informational cul-de-sac. Connectivity at a place like CERN was a little better. Here, we had a Local Area Network, or LAN, which was always on, and which allowed us to connect with any other computer in the complex. We had messaging systems, and better transfer speeds for files and technical information. We also all had specific user IDs, passwords and accounts, stored in each mainframe. Mine was 'timbl', a moniker I would use pretty much everywhere from then on.

By the mid-1980s, the big question in networking was how to get one LAN – say the one at CERN – to talk to another LAN – say the one at Stanford. The data would travel over the phone lines, or, later, by the first transatlantic fibre optic cable, which was laid in 1988. Once these systems were connected, they would cease to be 'local' networks and instead become a network of networks: the inter-net.

The debate was over what language, or 'protocol', these disconnected systems should use to talk to one another. American computer engineers working for the Pentagon in the late 1960s and early 1970s had laid the foundations for the internet by building the earliest hardware and making some of the first cross-network connections, but well into the 1980s there was disagreement over what the lingua franca of this global network would be. To my eyes, the best-designed protocol was something called TCP/IP (it combines the Transmission Control Protocol with the Internet Protocol). It was developed by two American network engineers, Vint Cerf and Bob Kahn, both of whom I would later have the pleasure of meeting. (Cerf and Kahn are sometimes called the 'fathers' of the internet.) Their protocol used 'packets', breaking information into small pieces at one end of the communication, then recombined them into something intelligible at the other end. Doing it this way was brilliant, since it didn't matter what order the packets arrived, or if one of them was for some reason delayed. Packet-switching, as it was called, relied on concepts from queuing theory, the same discipline my dad had demonstrated by lining us up behind a football when we were kids. The final result was an elegant and resilient method for transmitting information.

The US Department of Defense adopted IP addressing in 1981, making it the first early major user. For this reason, within CERN there was initially some consternation about the TCP/IP system – many Europeans associated it with the Pentagon, which was vilified in Europe, especially at that time. Several voices at CERN wanted to use a separate networking protocol called the Open Systems Interconnection, or OSI. This was being standardized by the International Standards Organization (ISO), the body in Geneva which had standardized a lot of things, including global shipping containers. In Europe, there was a strong political push for the ISO protocols. The disadvantage, which was major, was that it would create two separate systems that didn't speak the same native language.

This painful phase of internet history is known as the 'protocol war'. Fortunately, at CERN we had an inside man pushing for TCP/IP. Ben Segal was CERN's internet evangelist, and he waged a long and patient battle within the organization to abandon the international standard for the upstart American one – and he succeeded. Segal was intense, and very passionate about having one internet, rather than American and European versions which couldn't talk to one another – there were enough language problems already. He even brought in Vint Cerf to give a talk on the advantages of the IP system. He eventually won, and by 1988, everyone at CERN had an internet-facing email address. (OSI fell from favour and was eventually abandoned.)

Email was the internet's first killer app, and everyone at CERN used it. Another popular internet application was newsgroups, which served a need somewhat similar to the one Reddit does today, except it was decentralized – there was no central server. Another application was Telnet, which allowed users to access and control other computers remotely. Yet another application was File Transfer Protocol, which allowed users to send and receive data files remotely by accessing FTP 'servers' with a dedicated application called an FTP 'client'. They are so called because, just as a lawyer's client uses the lawyer's services, so the client of an FTP server uses the services of the FTP server.

Telnet, email, newsgroups, FTP – all ran on separate, dedicated applications. These programs weren't user-friendly, and FTP servers tended to organize information in a top-down, hierarchical way, rather than the path-independent approach the internet seemed to encourage. While spending my days programming various devices around CERN – and frequently stopping for coffee – I slowly began to conceptualize a different way of doing things. There was never one 'eureka' moment, just the slow and patient crystallization of an idea. Over time, this idea continued to fuel my persistence, and over time that persistence would turn into an invention that would change the world.

•

Before I describe the invention of the web, let me describe the work we were doing at CERN. Bear with me and you will see where this is going.

In the beam tube in the tunnel of the Large Electron–Positron Collider (LEP), a stream of electrons going one way would collide with a stream of positrons going the other. The electrons were made of matter, and the positrons were made of antimatter; when the streams collided, the detectors surrounding them lit up like a firework. What we were trying to engineer with these collisions were *new* particles, ones no human had ever observed. Unfortunately, we couldn't see these unstable new particles directly, as they disintegrated almost immediately. Instead, we used huge detectors – bigger than a house – to collect data about all the bits flying off. Looking at the pattern of those bits, we would then reverse-engineer a picture of what the original particle must have been. Within that picture there could be a massive discovery, one perhaps worthy of a Nobel Prize. There were four detectors, at four different places around the LEP tunnel. Each had been designed by different teams, each used completely different kinds of sensors to track the particles – and each was competing for that same Nobel Prize.

The problem was that each event picture was around 10 gigabytes large, and we were generating 40 million events a second – a mind-boggling amount of data. So the trick was to ignore almost all the pictures, but to have 'triggers' which would try to guess whether a particular collision event was potentially interesting. These triggers were arranged in stages. First – was there a big flash? If so, pass on to the next question, and so on. At each stage, the question about the event took more time to answer, but that was OK, because we'd screened out most of the events. At some point, after several qualifying stages, the much-reduced number of pictures would be sent to the surface computers for scientific review by a human.

The division I worked for was named Data and Documents. This division did a lot of the fun real-time stuff, but they also ran the computer centre, with giant mainframes manufactured by IBM, Control Data and Cray. The goal was to take all the data from the subatomic demolition derby we were running and turn it into something a human could read.

To present and format our findings, we used an in-house documentation system called CERNDoc. Now we arrive at the point of this story. Sometime in the late 1980s, I saw its author present CERNDoc to a small auditorium full of staff. To use his system, he explained, you had to write a file in the SGML markup language. Then you had to store it on the mainframe. Then you could categorize it, with a project name and a category. As the presentation continued, I saw how with each constraint the author of the software introduced, another part of the audience was alienated. 'What happens if the document is in Word?' one audience member asked. 'I want to store my document on my local computer, along with the rest of my life,' said another. 'In our division, we have our stuff arranged in a tree, not projects and categories,' said a third. And so on.

Watching the resistance to CERNDoc was an important lesson for me. If a documentation and storage system was to be acceptable to all the folks in that room, then it would have to make no demands of its users. It would have to impose no constraints. It would have to work with any format, any computer, any operating system, any categorization system and any sort of document. It would have to be a sort of *universal* space, in which all of their different systems could coexist. That universality would later be key to the web.

As a physicist, you are taught to generalize. An apple falls from a tree with a certain acceleration – would it fall from a hand the same way? What about a pear? What about a cannonball? In physics, you look for the most general rule, and if that holds true, then it will be the most useful rule to adopt. When you code, you also generalize. If you need a function to ask the user for their name and date of birth, you write a function that can ask *any* question, then you use *that* function to ask the name.

I began to wonder if you could generalize documentation systems in the same way. At that time, all documentation systems asked users to do some combination of reading information, selecting from a choice of options and typing in a search term. For Enquire, I'd permitted users to link any two resources together, but I'd still forced them to define what kind of resource each thing was. What if I didn't do that? What if I dropped all the categories, and let the user staple together any two things they could find? And not just any two things, but any number of things, with no restrictions on type, function or size?

In this way, I could build a new system – or rather a meta-system, a system of systems – that would be able to include a view of all the existing systems, but which could act too as a system for keeping track of my life and my projects. I guess I had two goals: one was that this should be acceptable and usable by anyone for anything. The other was that for running my own life, and managing my own collaborative projects, it should be a really powerful tool.

That seemed to be what the community at CERN was looking for. All these people could go on working just the way they always had, but the information they had at their disposal would be available to anyone on the internet. Behind the scenes, of course, you'd need a lot of computers and code, but all that could be hidden from the end user. Instead, the world's information would be split into two layers. There was the back-end layer, which stored and managed the information and the servers. On top of that was the client layer, which presented and edited that information in a manner that was as flexible and user-friendly as possible. In this way, I hoped to solve the lack of interoperability between document systems, but also have a collaborative tool which I would love to use myself.

But what would the user-friendly client layer look like? To me the answer was obvious: it should be made of 'hypertext'. First conceptualized in the 1960s, hypertext is a system for organizing and linking text in a non-linear way, allowing users to navigate through related pieces of information via 'hyperlinks'. Unlike traditional linear text, hypertext enables readers to jump from one document to another based on their interests or needs.

By the late 1980s, hypertext was a well-developed field. In a 1945 *Atlantic* article, 'As We May Think', Vannevar Bush, scientific adviser to the US president, had theorized about a kind of 'non-linear information retrieval' system that would allow users to create trails of links between documents, mimicking the associative processes of the human mind. Influenced by Bush's work, in 1963 the visionary entrepreneur Ted Nelson coined the terms 'hypertext' and 'hypermedia' as part of an ambitious computing platform he termed Project Xanadu. Nelson envisioned a universal, interconnected information space where every piece of content could be linked to every other piece, reflecting the non-linear nature of human thought. The system worked through a centralized system, which could ensure the integrity of all links and backlinks, to be run by the Xanadu Operating Company. (Nelson was never able to get Xanadu past the prototype stage.)

Building on these ideas, in 1968 the visionary computer scientist Douglas Engelbart gave a ninety-minute presentation at the San Francisco Civic Auditorium, which we now call 'The Mother of All Demos'. Remember, at this time, computers were gigantic $10 million mainframes that weighed several tons and mostly did scientific maths. They were not intended for regular office use, and many of them didn't even have monitors. Suddenly, Engelbart, wearing a headset microphone, pops up on a screen like some visitor from the future and introduces collaborative word processing via live teleconferencing feed while dragging a cursor around on his screen with a mouse. None of these technologies – not even the mouse – were familiar to anyone in the audience. In his presentation, Engelbart also debuted the 'hyperlink', which users clicked on with a mouse to move both within documents and between documents stored on a single computer. In fact, it was Engelbart who coined the term 'mouse'.

Hypertext gained widespread popularity in 1987, when Bill Atkinson, a programmer at Apple, released HyperCard, a 'stack' of virtual 'cards' that organized text, images and multimedia objects in a manner that resembles a modern web page. The cards weren't organized in a hierarchical way, and you could jump from one card to any other. (This was the kind of 'illogical' conceptual leap that computer programmers sometimes struggled with; Atkinson had conceived of HyperCard while on acid.) HyperCard was a massive hit. Apple bundled it as a default app with Macintosh computers, and a community of HyperCard enthusiasts grew around it.

•

So there is your potted history of hypertext – except at the time I was almost totally unaware of these developments! When I was conceptualizing the web as an information-wrangling tool, I had never heard of Ted Nelson nor Douglas Engelbart, and I had never used HyperCard. My introduction to hypertext only came while I was drafting a proposal for my document-linking system, when someone – I cannot remember who, but probably my CERN colleague Robert Cailliau – shared with me the July 1988 edition of the journal *Transactions on Computer Systems*, produced by the Association for Computing Machinery (ACM). Entitled *Hypertext on Hypertext*, the edition was electronic, delivered via floppy disk, and ran as its own executable program. After a snazzy splash screen, the newsletter displayed a table of contents, with each section highlighted. Click on a given highlight and you were immediately directed to that chapter of the document – just as you would be today from the contents page of an e-book. (This was not a new idea: Vannevar Bush, in 'As We May Think', had called for an electromechanical machine, the Memex, to do that with photocells and microfiche.)

By this time, many other documentation systems had links, such as links inside a document to a glossary. In Enquire, for example, after each note there were numbered links to other notes, a bit like references at the end of an academic paper. You could navigate to another note by specifying the number of the reference you wanted. But the hyperlink was obviously a more powerful solution, as it was more general. The really exciting thing was realizing that hypertext was a universal language in a sense: you could make the menus out of hypertext, and the documents out of it too, and you could end up mixing the navigation and the content.

Instead of shooting you just to the boring academic endnotes, it could shoot you anywhere else in the text. And soon I saw that the hyperlink could be even *more* powerful than its designers had realized. The problem with the ACM journal was that the hyperlinks were quarantined to one system, such as one CD-ROM. For whatever reason, the hypertext pioneers hadn't imagined that hyperlinks should jump *beyond* documents into any computer, anywhere in the world.

So that was it: my (simple, in fact) idea was that hyperlinks could provide a way for users to navigate the internet. You would click a link and it would take you to the next part of the document, or to a computer in the next room over, or to a server halfway around the world. Nor – and this was important – would you necessarily have any idea what was at the end of the link. It could be a photograph, a song recording, a diary or a manifesto. Other internet systems of the time made you specify what you were looking for. The link wouldn't do that. It would be a universal portal. Step through it and you could find yourself anywhere.

•

All of this was a pretty far cry from my day job programming the data bus that fed the detector images to the mainframe. CERN was focused on a specific kind of science, and while the place was unquestionably high-tech, the organization didn't have a formal mission to act as a technology incubator. To make my hyperlink concept a reality, I had to convince my bosses it was a good idea.

Fortunately, I had wonderful bosses. The Data Acquisition team was led by Peggie Rimmer. Peggie in turn worked for Mike Sendall, and Mike worked for David Williams, the head of the Data and Documents division. Peggie, Mike and David would later serve as my managerial protectors, directing resources and time towards the very first version of the web while deflecting criticism from budget-conscious overseers. You might call them the guardian angels of the web.

By 1989, I had begun to petition them to allow me to start the project. I had not really settled on a name for the idea, but in the memo I wrote to explain the idea, I included a diagram and titled it 'Mesh'.

Vague but exciting ...

CERN DD/OC | **Tim Berners-Lee, CERN/DD**

Information Management: A Proposal | **March 1989**

Information Management: A Proposal

Abstract

This proposal concerns the management of general information about accelerators and experiments at CERN. It discusses the problems of loss of information about complex evolving systems and derives a solution based on a distributed hypertext sytstem.

Keywords: Hypertext, Computer conferencing, Document retrieval, Information management, Project control

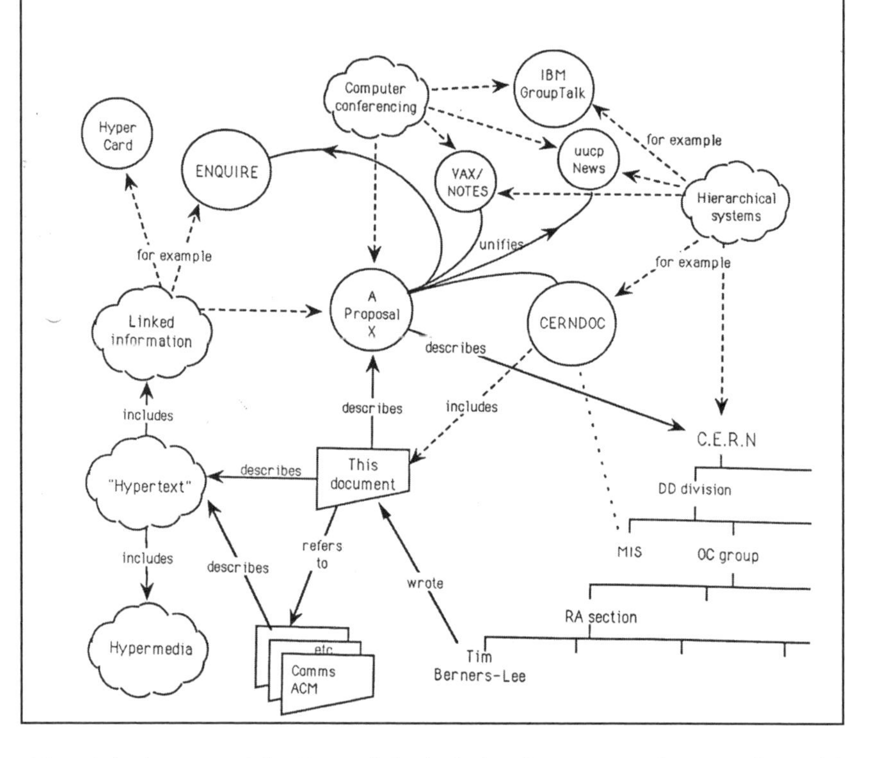

My original proposal for the web included a diagram, so that people could immediately picture how 'Mesh' allowed for connections and relationships.

This diagram was a concept graph, which uses nodes, edges and labels to describe entities and the relationships between them. In the centre of the diagram was a circle for the proposal itself, called 'Mesh'. On the right were arrows to concepts like 'hypertext' and 'linked information' in cloud-shaped bubbles. Above were circles representing existing systems like 'notes' that would be unified within Mesh. On the right was a little bit of an org chart showing how the author, me, is related to CERN.

The diagram itself is not the main point of the argument: in the memo, it was just a cover picture. If you like, the cover diagram was an alternative way of approaching the topic, for someone who could grok – intuitively understand – circles and typed arrows instead of natural language. In fact, the tension and complementarity of these two different ways of looking at the world, is a recurring theme in my work.

The accompanying note talked about a possible *global* hypertext system, without naming it. The problem of managing this sprawling information, I intuited, would soon be a widespread one. 'CERN is a model in miniature of the rest of the world in a few years' time,' I wrote. 'CERN meets now some problems which the rest of the world will have to face soon.'

The original memo was dated March 1989. Nobody jumped on it and insisted on picking it up. To a certain extent, that was reasonable, because CERN didn't have a process for picking software projects to do – it had a system for picking high-energy physics experiments to do.

A while later, someone – maybe David, maybe Mike – asked me, 'What happened to that hypertext thing you were going on about?'

'Nothing!' I said.

'You should write a memo about it,' whoever it was said.

'I did.'

'Did you send a copy to me?'

'I did.'

'Could you send it again?'

'Sure.'

So I reprinted the same memo, adding the new date as well as the old one ('March 1989, May 1990'), and sent it around again. Along with my concept graph, the Mesh proposal now included the following text:

> CERN is a wonderful organization. It involves several thousand people, many of them very creative, all working toward common goals. Although they are nominally organized into a hierarchical management structure, this does not constrain the way people will communicate, and share information, equipment and software across groups.
>
> The actual observed working structure of the organization is a multiply connected 'web' whose interconnections evolve with time. In this environment, a new person arriving, or someone taking on a new task, is normally given a few hints as to who would be useful people to talk to. Information about what facilities exist and how to find out about them travels in the corridor gossip and occasional newsletters, and the details about what is required to be done spread in a similar way.

This was my take on life, people and organizations. The regimented structure of the directory tree, employed in systems like the Unix file system, didn't replicate the actual flow of information in person-to-person systems. Chance encounters and happenstance played a critical role. For many computer scientists, designing things this way was 'illogical', and every document belonged in a specified container. I was proposing instead to free those documents – essentially to dump the files from their folders onto the floor.

What I was beginning to see was that information was meaningless in isolation. Instead, what truly mattered was the *relationship* between one piece of information and the next. Context was everything, but a hierarchical structure 'organized' information by quarantining it, which actually reduced its value. What you wanted, instead, was to encourage new and unexpected relationships between pieces of information to flourish. And, to do that, you had to let the *users* make those connections, in any way they saw fit.

I began talking about Mesh non-stop. My excitement was uncontainable. Most ideas don't scale; at a certain point, they hit limits of cost or physical practicality. But the connected 'web' I envisioned seemed limited only by imagination. The more links you made, the better it would become. The more it grew, the better it functioned. By layering hypertext links onto the internet, we could connect scientists and artists and citizens all over the world. I started to talk about the 'two Cs' – creativity and collaboration. By combining the two Cs, you could reach a new state of intellectual production I termed 'intercreativity': the ability of a *group* of people to be creative.

Almost no one got it. I will admit that, when I get very enthusiastic about something, my mind begins to race, and I can't talk fast enough to put my thoughts into words. I also have a tendency to explain things at the maximum level of abstraction, focusing on the broadest possible applications rather than specific examples. (Both my parents, I must remind you, were mathematicians.) Sometimes, in meetings, my colleagues would literally hold up signs reading 'Tim, slow down.' In a recollection written in 2014, Peggie Rimmer remembered my efforts to promote Mesh:

> [Tim] was obviously a very smart young man (smart-clever rather than smart-sartorial!), full of fizz and, as a bonus, entirely likeable. When he presented his ideas in our section meetings, few of us if any could understand what he was talking about. . . . We sometimes asked him to put things in writing, which didn't necessarily help either. One of his erstwhile colleagues recalls 'we knew it was probably exciting, maybe even important, but that it could take hours to figure out'. . . . At least I don't recall having to field any complaints, apart from 'what on earth is Tim proposing?'

Mike knew he would not get budgetary support for me to work on Mesh directly – so he thought of a way to do it indirectly. In the Data and Documents division (which would later, more accurately, be renamed Computing and Networking), we looked at computing in general, and we had people trying out the various different types of minicomputer and workstation. We had Digital Equipment Corporation's VAX systems, we had Hewlett Packard machines, we had Apollo workstations, and so on. But what we didn't have was Steve Jobs's new NeXT computer. Mike proposed we purchase one and try it out.

In 1985, Jobs had been forced out of Apple in a boardroom coup. NeXT was his follow-up company, free of all of Apple's history, and Steve Jobs used that opportunity to shoot for the moon. Among many other neat features, the NeXT computer was said to have great development tools, so if one were to want to write a new app from scratch, the NeXT would be a good machine to use. Mike said he'd buy me one so I could kick the tyres – check it out, in a way – on behalf of others in the CERN community.

'Once you get it, why not try programming your hypertext thing on it?' Mike said. I noticed a twinkle in his eye as he said it.

Jobs's agreement with Apple barred him from building a competing personal computer, so NeXT had moved into the professional workstation market. The model I received in 1990, the NeXTcube, retailed for $7,995 – more than $18,000 today. The computer, a one-foot die-cast magnesium cube, looked like a piece of hardware lifted from a spaceship. It came with a clunky black CRT (cathode-ray tube) monitor, a black rectangular two-button mouse and a black clackety-clack keyboard. When you turned on the computer for the first time, you got a pre-recorded audio message from Jobs himself. Riffing on the concept of the personal computer, Jobs described the NeXT as 'the world's first *inter*personal computer'. I felt as if he was speaking to me personally.

The NeXT computer I used at CERN – a crucial tool in the creation of early web software.

The NeXTcube was very cool in many ways. It used read–write optical discs, similar to CDs. It had a large, beautiful, 1024 x 1024-pixel greyscale display. It used the printer language Postscript as its display language, so its typography was immaculate. It ran Unix, my favourite operating system. It had a digital signal processor so it could manipulate speech and music. It was the Harley-Davidson of computers.

The NeXTcube also had this great development environment. It had code-writing and debugging tools, and it had an Interface Builder program which allowed you to make new applications quickly, by dragging and dropping the components of the app together before you even wrote any code. And it came packaged with a number of useful pre-existing code libraries.

Noticing how well Jobs marketed his products, I began to reconsider my own project's name; 'Mesh' didn't seem to convey the scope of what I was working on. (Also, it sounded like 'mess', which was perhaps a little too accurate.) I decided my project needed a catchy acronym. Some names I considered, but rejected, included the Mine of Information (MOI) or The Information Mine (TIM). I ultimately decided to call my system the 'World Wide Web'.

Many people at CERN grumbled at this name and complained about the nine-syllable 'WWW' acronym even more. Who makes an acronym – a shortening – that takes longer to say than the original name? (Today, when I tell students the story of picking the name, I tell them I Googled the term and found no one else was using it – then I watch their faces to see if anyone gets the joke.) But if 'World Wide Web' was wonky, it was also memorable. And it was practical, as my unique code modules starting with 'WWW' would not bump into other people's work. Also, when you used it as a domain name, like www.cern.ch, it would be clear the computer was probably running a web server – so we would be able roughly to count the growth of the web in terms of numbers of servers. And it meant a global graph, a *web*, which is what it was: a web that did not have a single trunk, in the way a tree did.

CHAPTER 3

Ignition

From around October 1990, I became the world's first full-time web developer. Mike and Peggie had helped me justify the work to CERN as an experiment in the NeXT programming environment. Writing in Objective-C, NeXT's default programming language, I began to code the software the World Wide Web would need to work.

How do you program a link that can lead anywhere? It's easier than you might think: you only have to build the door, not what's on the other side. Before the web, the preferred method for getting information from another computer was the File Transfer Protocol. There was a lot of stuff available on FTP servers, from open-source code to online books to visual art. People with software projects would run an FTP server – say ftp.acme.com – then encourage people to get the software by saying something like 'log into ftp.acme.com and then go to directory pub/code, and then get sparkle.c'.

But this method was inherently slow. You had to open a control connection to a server on another computer, then select the directory on the server you'd reached, then say you wanted to get a file, and then the client would open a separate connection to the server. Each connection would take a certain amount of time, depending on how far across the world the server was. It also demanded multiple inputs from the user – logins, passwords, directory navigation and downloads. Obtaining a file from an FTP server for the first time often took about a minute. With the web, I was aiming for a tenth of a second.

To do so, I needed the hyperlink to execute a command similar to an FTP request instantaneously, at the click of a button, and to do it in a way that was hidden from the end user. So I designed a new protocol, similar to FTP, but engineered to deliver the result as fast and seamlessly as possible. I called it the Hypertext Transfer Protocol, or HTTP.

HTTP was a new way for computers to talk to each other, but I deliberately kept most of what it did hidden from the user. In the original specification for HTTP, I called the hyperlinks 'Universal Document Identifiers', or UDIs, but for reasons I'll explain later people today call them Uniform Resource Locators, or URLs. In the interest of simplicity, I will use the term URL in this book, but let me stress that the original term *universal* was very important to me. Universal meant that, in addition to file transfers, we could later use these links for all kinds of other things. Universal meant that whatever wild and wonderful system someone down the road might design, I could just make a prefix for it and then map all the new parameters into the identifier. Universal meant an ever-evolving set of instructions to handle not just documents, but audio, video, and even interactive applications like maps, video games, online forms and automatically generated worlds – literally *anything*.

The URLs were in fact the most crucial, most innovative piece of the web. The idea that in a single short identifier you could encode all you needed to link to anything, and that the identifier space was unlimited, was the core thing. I cannot emphasize enough how different this was from all existing internet communications protocols at the time, which very much wanted to direct users towards *specific* kinds of information. By contrast, I believed there needed to be a universal naming system for the internet.

In fact, the URLs worked at such a high level of abstraction that it was often difficult to explain what they did, even to technical people.

'What's at the other end of the link?' they would ask.

'Anything,' I would say.

'OK, but what specifically?' they would ask.

'That's exactly the wrong question to ask,' I'd reply. 'Think of anything. The link will take you there.'

This brief snippet of conversation was repeated a great many times with a great many puzzled interlocutors. Since the concept of a truly *universal* system was baffling to people, I decided to make the rest of the hypertext protocol function in paradigms computer scientists were familiar with. In fact, the entire design of the web was all about using formats and languages and concepts which people might have come across elsewhere.

I began with the naming system. Single-document hypertext destinations had names like 'TableOfContents' or 'Section1', but for a sprawling network like the web we needed more specific identifiers.

The first step was to allow a link to include a file name, so it could take you anywhere on the computer. Instead of just 'Section1', the destination might read:

/users/alice/book/chapter3#section1

See the hash sign, and the computer file name (i.e. /users/alice/book/chapter3) before it? To the right are identifiers for objects in the world of hypertext, like link destinations, and parts of a document (i.e. section1). To the left are the bits about networking. In the marriage of networking and hypertext that is the web, that hash sign is where they meet. My idea was that people would develop all kinds of cool ideas around hypertext, which would only affect the bit on the right. Other people would develop all kinds of cool ideas around networking, which would happen on the left. And the hash sign would sit in the middle. If you learn anything about the design of the web, let it be this: the design merges networking and hypertext, and they meet at the hash sign. If you get that, then you basically get it.

But we needed to go further than just the computer. We needed addresses for the *network*. The high-performance Apollo/Domain workstations of the time had an elegant naming scheme that at least some people would have come across, so I used it as an addressing system for the web. Apollo workstations started addresses with two slashes, like this:

//alicescomputer/users/alice/profile

I borrowed this for HTTP, leading to a URL that might look like this if you were trying to retrieve a document called Sparkle:

http://www.acme.com/pub/code/sparkle.html

or for Section 1 of that document:

http://www.acme.com/pub/code/sparkle.html#section1

The code for the URLs was small, but it was the most important thing I ever wrote. Before the URL existed, you had to type in a command and hit 'Enter' to jump from one computer to the next. After the URL took over, all you had to do was click – or later touch, or tap, or speak, or merely gesture. The URL reduced the friction in internet navigation to zero. As telecommunications throughput improved, it also made jumping from one server to the next an invisible, near-instantaneous process. Many younger readers, I suspect, will be baffled that there was ever a time the internet *didn't* use hyperlinks. When an invention becomes truly ubiquitous, it can seem invisible.

•

A critical component of HTTP was the server-naming system. Again, wanting to give users familiar paradigms, I opted to piggyback on the pre-existing Domain Name System (DNS). In Cerf and Kahn's internet, each computer is given an IP address – for example, 185.249.56.23. In the early 1980s, engineers decided it might be nice to have memorable human names to go along with these numerical addresses. The resulting DNS protocols permitted organizations and individuals to register such names within specific, delineated 'domains'. Military users were given the domain '.mil', governments were given '.gov' and educational institutions were given '.edu'. Businesses, of course, got '.com'. These addresses defaulted to the US; other countries had to use country-specific domains like '.co.uk'. The first DNS registrant, in 1985, was 'symbolics.com'. This name pre-dates the launch of the web by about five years.

Later, these two technologies – DNS and HTTP – would become conflated, but I think it's important to be clear about what I invented (URLs, HTTP and HTML) and what I did not (the dot-com names). What I did do was to harness the power of the DNS approach, so that if the (now-defunct) Symbolics Corporation of Massachusetts wanted to host a web page, it could do so with the name http://www.symbolics.com/sparkle.html.

When you click on this link, your computer looks at the URL, saves any '#section1' bit for later, and takes up the first bit, 'http:'. This tells it to prepare to initiate a connection to another computer, using the hypertext protocol. Then it takes the next bit and opens an internet connection to the other computer, whose name is www.symbolics.com. Then it sends the rest of the URL, sparkle.html, to that computer, and that computer delivers back the contents of the web page – going to Section 1 if there was that bit after the hash sign.

This is where the third part of the design of the web comes in, as the web page has to be encoded in some language. At the time, there was no standard format for hypertext, so I defined one: the Hypertext Markup Language, or HTML. Again, the rule for designing the web was to use existing stuff wherever possible. Much of the original version of HTML was based on Standard Generalized Markup Language, or SGML – which was being used at CERN already for the documentation systems on the mainframe. SGML was a bit clunky in a way, but that was fine with me – I didn't intend for users or even creators to ever see HTML, any more than users of a word processor would look at the invisible markup language used to format files for the printer.

I viewed HTML more like plumbing. HTML rendered pages using elements called tags. So, for example, you separated the title for your web page from the body of text, with the 'title' and 'body' tags, like so:

```
<title>Welcome to the web</title>
<body>The web is for everyone!</body>
```

This would then render as:

Welcome to the web
The web is for everyone!

The original version of HTML was quite simple. It had headings and lists, but no inline images. Instead, you could link to images, which would pop up in a different window. Of course, HTML has evolved over time; we're now on Version 5, and modern HTML is more like executable software, with limitless capabilities. Still, the formatting remains the same. That simple core of the HTML language is *still* the core of HTML, and web pages made with it will still work well in modern browsers.

•

In addition to defining the protocols for the web, I developed client and server software. The server was the package that would 'host' the web pages, which would be located on internet-enabled computers. Users would navigate between these servers to get from one web page to the next. Although this software would mostly operate outside the view of the public, it was critical for developers. In the same way that a fancy retail store might have an unglamorous warehouse and distribution network supporting it, the web server handles all the requests from users and does all the behind-the-scenes technical work.

The last – and, for end users, most familiar – application I coded was the web 'client'. This was the program that rendered the pages written in HTML. My first client, which I called simply WorldWideWeb.app, was a read–write tool that allowed users not just to view web pages, but also to create their own. In my original vision, the web client allowed users to edit web pages that already existed – sort of like Wikipedia today. As it turned out, users preferred to use one application to read web pages, and another application to write them. This rendered the web a lot more passive than I intended it – but since such applications are primarily used for reading, today we would call my web client a 'browser'.

I had a lot of fun coding the WorldWideWeb application. There was definitely the zen-like focus of 'being in the zone', and the creative thrill of making a new thing from your imagination. People call it getting lost, but actually, you don't get lost at all. It's more like being very *found*.

I started by firing up the development tools which NeXT was famous for. Select 'New Project'. Name? 'WorldWideWeb'. Add a window for the app to put the web page in. Add a panel for the user to be able to inspect the details of a document or a link. Then control–drag a menu item into the panel, so that when you click on the menu item, the panel will appear. The beauty of the NeXT system (thank you, Steve Jobs!) was that a lot of the application was built just by dragging and dropping. In the panel I could put buttons, and then experiment with what would happen if someone pressed the button. In some cases, it would call for a new bit of code I wrote from scratch.

The final source code for the first version of the web consisted of 9,555 lines of Objective-C code. Given what I was trying to achieve, this was really a compact amount of code – and that was because of NeXT. In fact, NeXT's Interface Builder was what made the web possible. Using its elegant, drag-and-drop development environment, I was able to do in a few months what would have taken years in Windows. Many years later, we used blockchain technology to package this original source code into an NFT (non-fungible token), which was then auctioned off by Sotheby's. The NFT of the code was compressed onto a big poster.

Another reason I developed the web so quickly was that I had a deadline: our first child was due on 24 December, Christmas Eve! The excitement of developing the web was totally overshadowed by the coming baby. CERN was also shutting down in the middle of December, to save energy. The offices and all of the physics would be turned off, so I had to wrap up the project in a state in which I could leave it for a couple of weeks. I added a release date as a version number: 19901225 – Christmas Day. This was actually a bit of a fib: the app was finished by the 13th, but I figured

Christmas Day would be more memorable. (To this day, some people assume I was working at CERN till the 25th in an effort to get the code out.)

As it turned out, I didn't have to rush: my daughter Alice kept us waiting. She was born on 1 January, and was in the Swiss papers as one of the first babies of the year. It was a totally brilliant and wonderful start to 1991. My happiness was doubled when my son Ben was born in 1994. When people ask me about this time, I don't think about the web; I think about little Alice, determinedly dressing herself in the morning, and later of little Ben, just learning to crawl. To this day, I keep track of my life not by calendar time, but by associating dates with the stages of development of my two children.

•

A number of decisions I made in that first year of web design had far-reaching consequences I didn't intend. First, by relying on the domain name system I ended up tying the web to the instability of domains. There was an alternative to DNS which the UK had been using, where the parts were largest first, like 'com.microsoft.www' rather than 'www.microsoft.com' – but that was not what the world was adopting. So I chose the American way, although ultimately the British way would have made more sense. Here's what I mean: logically, web addresses should become more specific as you go on, so it should really read ch.cern.info, since 'ch' – the country code for Switzerland – is more general than CERN, the research facility in Switzerland. Under this scheme, a business address might logically read, uk.co.widgets.www, specifying that we are going from the UK, to a company, which sells widgets, and then to its web server, instead of the www.widgets.co.uk, which we have now.

Now, at the time I was making these decisions, DNS was run in the public interest. But, although I could not have foreseen this at the time, the administration of the DNS later fell into the hands of companies who auctioned off popular domain names, like real estate. During the dot-com boom, those operators turned the supposedly non-profit DNS address system into a nauseating free-for-all. Domain names would become an asset class for investors, rather than a tool for innovators, and this led to the practice of domain-name squatting, or 'cybersquatting', where unscrupulous speculators registered popular top-level 'dot-com' addresses, in the hope of flipping them for a profit. More on this later.

As the web evolved, I was sometimes surprised by the turns it took. For example, I never really meant for users to see HTML – all of that was supposed to be done by the read–write web client tool. There had been a time when people would use markup in the documents (writing '.bold' on a new line for bold, for example) but by 1990 people were used to early word processors, like MacWrite or WordStar, which let you see the formatted text immediately as you wrote. These were called 'What You See Is What You Get' editors, or WYSIWYG for short. To my surprise, though, users were keenly interested in HTML and were prepared to code it by hand, with all the tags. Many successful web pages were first coded in raw HTML, and learning to write HTML became a rite of passage for the first generation of web publishers.

I wanted users to browse the web from link to link without having to be aware of the underlying technology. That is, until something went wrong; then the browser had to explain what had happened. So I programmed a set of numbered error conditions in HTTP, which were copied from the error codes of other protocols. The 400-level status codes meant the user had done something wrong; 500-level status codes meant the problem was on the server side. The most famous of these is error 404, which indicates the URL you are attempting to access doesn't exist, or isn't available. As the web evolved, so did the number of 'dead' hyperlinks, causing the 404 error to become a bit of a meme. The number of pages which have been allowed to rot, or have been deliberately broken in a 'wonderful new' website design, is shocking.

Lastly, using the rather obscure Apollo naming system meant placing the two slash characters '//' between the http protocol specification and the DNS address. The idea was that if you were making a link between two websites, then you could skip the 'http:' and just start with '//' – but in practice people don't. So I could have designed the web without the double slash, and that would have led, over three decades, to quite a lot of saved keystrokes. Sorry!

Then again, maybe early adopters, people who had seen other systems with double slashes in them, would not have picked it up, and we would not be using the web today. Who can say?

•

In 1989, I had explained my vision for the World Wide Web to Robert Cailliau, my colleague at CERN. Robert was a sharply dressed Belgian who absolutely worshipped at the Cult of Mac. He got what I was trying to do at once, and became an immediate convert.

Robert's evangelism for the web rivalled my own, and in 1990, he managed to convince his boss to buy him a NeXT computer, too.

By December 1990, I was hosting the world's first web page on the NeXTcube in my office – if you turned off this machine, you would turn off the World Wide Web. I later pasted a sticker to the front of the machine, and wrote, in red marker, 'This machine is a server. DO NOT POWER IT DOWN!!'

If I had walked to Robert's office, booted up my WorldWideWeb.app program on his NeXT computer at the time and had typed in http://info.cern.ch/hypertext/WWW/TheProject.html, then this is what we would have seen:

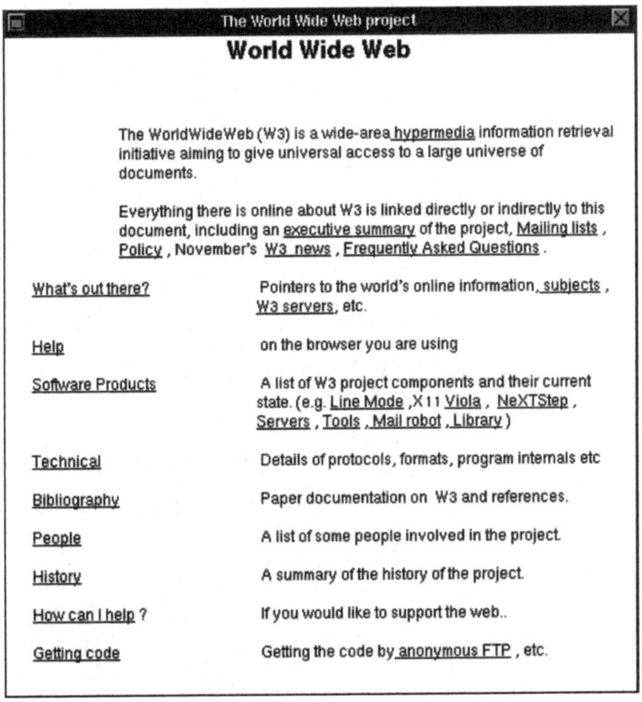

The World Wide Web project

World Wide Web

The WorldWideWeb (W3) is a wide-area hypermedia information retrieval initiative aiming to give universal access to a large universe of documents.

Everything there is online about W3 is linked directly or indirectly to this document, including an executive summary of the project, Mailing lists , Policy , November's W3 news , Frequently Asked Questions .

What's out there?	Pointers to the world's online information, subjects , W3 servers, etc.
Help	on the browser you are using
Software Products	A list of W3 project components and their current state. (e.g. Line Mode ,X11 Viola , NeXTStep , Servers , Tools , Mail robot , Library)
Technical	Details of protocols, formats, program internals etc
Bibliography	Paper documentation on W3 and references.
People	A list of some people involved in the project.
History	A summary of the history of the project.
How can I help ?	If you would like to support the web..
Getting code	Getting the code by anonymous FTP , etc.

The first web page!

This was the first web page, designed to help people learn about the web and get on board. There was only one colour – the NexT was a monochrome machine. There were no images embedded in the web page, although there were images on the web that you could link to. You had to use a little imagination to see the web's full potential, but it worked.

The rest is history. Except that when you write about it many years later, you learn some things which you never understood at the time. Famously, when Mike died ten years later, Peggie, his wife, obtained his original copy of my memo, with the words 'vague, but exciting' written in pencil on the first page. They had been kept hidden from me all those years. I later discovered that during that time Mike and Peggie were facing some profound personal upheaval. Mike was wondering if he might be gay. Society was not so understanding of sexuality at that time, and Mike was unsure what to do with that side of his life. Also, around then, Mike was diagnosed with cancer and given between eighteen and twenty-four months to live. So, to put it mildly, it is little wonder that he wasn't getting too involved with things like my hypertext project. And yet, even through this difficult period, I had a huge amount of support from both of them.

Peggie and Mike had been partners in a relationship which seemed to work well for them: him living in chaos in France, she living in order in Switzerland – at work keeping separate, but in their time off going on holidays together. They were opposites in personality, she being expressive and involved, and he being rather more introspective and withdrawn.

Peggie reoriented her life to be able to help Mike, and despite the apparently irreconcilable problem of his sexuality, getting married was suddenly a necessity so that Peggie could care for Mike in a Swiss hospital. Mike had ten years in the end, rather than eighteen to twenty-four months. He died in London, 15 July 1999. He was a lovely, wise, kind, funny, clever person. On his office door, I remember a *Punch* cartoon of the Daleks from *Doctor Who*, invading the world on their castors, before encountering an unforeseen obstacle.

"Well, this certainly buggers our plan to conquer the Universe."

The Punch *cartoon on Mike Sendall's office door.*

Mike truly played a key role in helping the web to conquer the universe. We miss him.

•

The environment at CERN was also important to the development of the web. As I'd observed in the memo, people at CERN tended to resist being hierarchically organized; they preferred to make unexpected lateral connections without reference to job title or divisional structure. There was also something of a 'Medici effect' happening there, with brilliant people from many disciplines and nationalities converging in a single place. A conversation at CERN could lead anywhere, from a discussion of Rutherford's early particle physics experiments, to the merits of various architectural approaches in computing, to polyglot puns and wordplay in different languages, to downhill skiing techniques, to the latest developments in Eastern Europe and the collapsing USSR, to the best place to get an espresso in Milan. From these seemingly random topics of conversation a great many unexpected connections could be made.

I cannot overstate the importance of these unexpected connections. In the modernist era, urban planners like Le Corbusier designed 'rational' cities, which segmented neighbourhoods by function and stripped buildings of detail and ornamentation. In so doing, they did irreparable damage to the human spirit. I explicitly conceived of the web to be fractal, thumbing my nose at this kind of false 'rationality'. The result – very much by design – was, in Peggie's words, an 'anarchic jumble' from which anything might emerge. And this worked much better than anyone anticipated, precisely because I didn't want to enforce any particular structure, so things on the web took any shape and size.

Behind the scenes of the web, of course, there are computers, and those computers are distinctly modernist creations. They have to follow rules, they have to be segmented by function, and they have to be utterly rational. (Similarly, behind the scenes in cities, we might prefer orderly systems of electricity, sewage and transport.)

But too often, software designers, like urban planners, make the mistake of demanding that the user – the human – acts like the system, rather than the other way around.

The key thing to consider was the *links*. Humans are social animals; a person raised in isolation has no identity, no culture, no voice, nothing. It is our links with other humans that define who we are. Those links take the form of language, of culture, of sexuality and family and religion and ideology and productive enterprise. From each human, a sprawling network of individual links connects them to their friends, their family, their society, their culture and the planet.

For much of human history, those links remained fairly fixed; most people were bound by narrow limits of ethnicity, religion and location. But as our species progressed, technology made new links possible, connecting us across cultures, oceans and time. By the late twentieth century, the average citizen of this planet was connected in a sprawling network of trade, culture and communication to billions of other citizens around the globe.

With the web, we were at the outset of something major, and we had to design it with the human being first in mind. We had to build a system that gave humans the ability to make links around the world, but one that avoided ensnaring them in dead-end, anti-human materialism, or systems of surveillance, coercion and control. Properly designed, such a system could multiply the links that define us by many orders of magnitude, and open up a new era of creative potential. Computers had never modelled those links before; it just wasn't the way most computer scientists thought about things. But my background and upbringing suggested to me that there was no technical barrier to building such a system. This, above all, was what got me so excited about the World Wide Web.

•

Even with a functioning website, it was still something of a struggle to attract users to the technology. The biggest problem was that my web client only worked on NeXT computers. There were only a dozen of those at CERN, and only about 50,000 worldwide. If I was going to enable a new era of human creativity, I needed to make my system more accessible. 'This is not just for scientists, not just for academics,' I said to myself. 'This is for everyone.'

I recruited a young CERN intern named Nicola Pellow to write a universal web client. By spring 1991, Nicola had created a beta version of the command-line program that would work on practically any computer, and we released it to a limited audience within CERN. Nicola's program was, in some ways, the first true 'browser', insofar as it was read-only, and platform-independent. When you navigated to info.cern.ch using the Telnet program, your screen would show the image on the page opposite.

Primitive, yes – you couldn't even point and click! Instead, the links were 'footnoted' and you navigated by entering the appropriate number into the command line. The limitations of Nicola's browser were obvious, but the advantages were that it could be used on any computer at all, even a paper Teletype. From the start, universal access was part of the web's core philosophy.

I began to distribute Nicola's browser to my colleagues at CERN, but I soon realized the browser was useless without something interesting to direct it to. It was the classic chicken-and-egg problem. Without browsers being available, no one would be motivated to build a website. Without websites being available, no one would be motivated to make a browser for different systems. In a way, the line-mode browser, though simple, left it so that no one had the excuse that they could not browse the web.

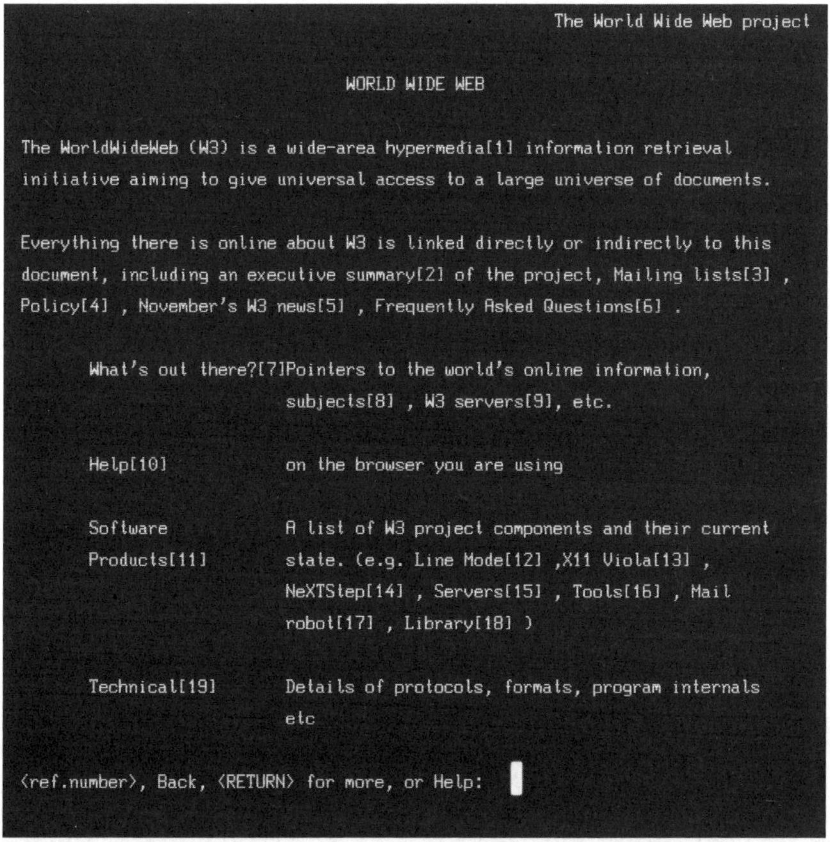

```
                                                    The World Wide Web project

                              WORLD WIDE WEB

The WorldWideWeb (W3) is a wide-area hypermedia[1] information retrieval
initiative aiming to give universal access to a large universe of documents.

Everything there is online about W3 is linked directly or indirectly to this
document, including an executive summary[2] of the project, Mailing lists[3] ,
Policy[4] , November's W3 news[5] , Frequently Asked Questions[6] .

     What's out there?[7]Pointers to the world's online information,
                     subjects[8] , W3 servers[9], etc.

     Help[10]            on the browser you are using

     Software            A list of W3 project components and their current
     Products[11]        state. (e.g. Line Mode[12] ,X11 Viola[13] ,
                         NeXTStep[14] , Servers[15] , Tools[16] , Mail
                         robot[17] , Library[18] )

     Technical[19]       Details of protocols, formats, program internals
                         etc

<ref.number>, Back, <RETURN> for more, or Help:
```

A screenshot of the universal web client – the command-line program could run on almost any computer.

We also needed a website! Meanwhile, CERN needed a better phone book. The CERN phone book was online, but you had to log onto a big IBM mainframe to use it. People who never used the mainframe for anything else were keeping an account on there just to be able to look up phone numbers. Bernd Pollerman, the person running the phone book, was charged with finding a way of getting people on other platforms access to it, and together we cooked up a scheme where he would run a phone book web server on the mainframe, and then people would access it using web browsers. Dietrich Weigant, the person running the big Unix machines, installed the line-mode browser, so you could type 'www' to get the web, or 'www http://cernvm.ch/phone' to get the phone book.

This proved popular, and so, from the CERN point of view, the phone book was the killer app for the web. It gave everyone at CERN an incentive to install a web browser, although it did have the unintended effect of encouraging users to be passive consumers of content, rather than active creators. By mid-1991, about a thousand people at CERN were regularly 'using' the web, but mostly to look up phone numbers. This rather undersold the potential of the technology. Curious to see if the web was spreading outside my institution, I set up a server log to track how many outside visits info.cern.ch received. By August 1991, I was receiving more than ten page views a day.

•

In late 1991, I travelled with Robert Cailliau to attend the Hypertext '91 conference at a hotel in San Antonio, Texas. I brought with me my NeXT computer and a modem, intending to wow the conference attendees (including Douglas Engelbart) by remotely logging into my web server in Switzerland. But no one at the conference had an internet connection! Now, the hypertext community had all sorts of great ideas: video games and digital magazines and CD-ROM encyclopaedias. For whatever reason, no one had thought of using the hyperlink to jump out of one computer and into the next – not even Engelbart.

Internet connection or not, Robert and I had to show them. Robert convinced the hotel manager to string a phone line from the lobby to the conference room, allowing us to hook up the modem. He then called the nearby University of Texas at San Antonio and persuaded someone there to let us use their dial-up connection. Finally, he jerry-rigged our modem to the American 110-volt power supply and the US phone line connection with a soldering gun. In this MacGyver-like fashion we were able to demonstrate the web to the conference attendees.

The demonstration was a hit, but the World Wide Web project remained precarious. There was an unspoken question that haunted the project, hinted at but never precisely articulated: 'Why, exactly, is the European taxpayer financing the oddball passion project of a thirty-six-year-old computer programmer in the Data and Documents division of a Franco-Swiss particle physics laboratory?' Peggie and Mike were my guardians, but at any time their superiors could have directed me to stop wasting time developing internet communications software, which was neither part of my job description nor part of CERN's mission.

It was true that CERN was generating just a surreal amount of data, and it was true that many people wanted to access that data online – but you really had to squint a little to see the connection between particle physics experiments and the World Wide Web. To justify my efforts, I had to accrue a critical mass of both users and content developers. At the end of 1990, I was basically the only person who had ever used the World Wide Web. By the end of 1991, in addition to the phone book lookups, I was getting 100 hits a day on the CERN server. I never assumed that this exponential rate of growth would continue, guaranteeing 1,000 hits per day by the end of 1992 and 10,000 per day by the end of 1993. Yet, that is exactly what happened.

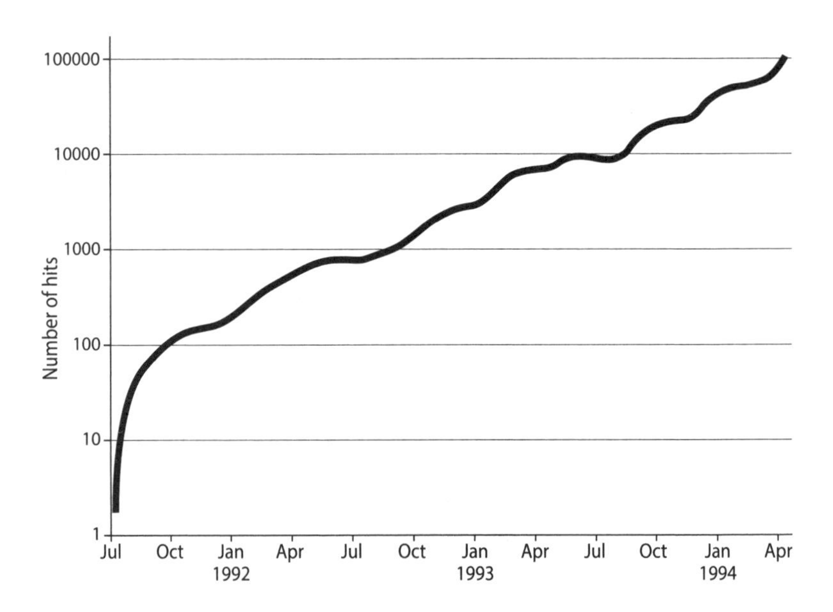

The load on the first web server in hits per day across three years, showing exponential growth.

Little did I know that there would be a billion global web users soon after the turn of the millennium, continuing the exponential trend. The growth still continues to amaze me – there are now over 5.5 billion users of the web, or close to 70 per cent of the world. But never at any stage could I relax and assume that the growth would continue.

CHAPTER 4

The Rise of the Web

Estimates vary, but only around 2 million people regularly used the internet in 1991 – approximately 0.03 per cent of the Earth's population. Most of them were academics, since the internet hadn't yet been formally opened to the public, and the concept of an Internet Service Provider (ISP) was still in its infancy. One of those early users was Paul Kunz, a particle physicist and software developer at the Stanford Linear Accelerator Center (SLAC) in Menlo, California.

Paul visited CERN for unrelated business in 1991, but, as I knew he was a NeXT enthusiast, I was able to corral him for a quick demonstration of the World Wide Web. Paul used a NeXT at CERN to log into his own computer in California remotely, then downloaded the web app remotely. He loved how fast it was, even when browsing web pages across the Atlantic. Like Robert Cailliau, Paul got it instantly, and evangelized for the technology upon his return to Stanford. With the help of SLAC librarian Louise Addis, Paul encoded a large number of scientific papers into HTML, then created a web page at http://slacvm.slac.stanford.edu. This, the first website hosted outside of CERN, proved a hit, and, by the end of the year, researchers at Stanford were using it regularly to search SLAC's archives. The SLAC site proved not only that the web was useful, but that it was useful for high-energy physicists in particular. After that, there was a lot less tension at CERN.

Meanwhile, I continued to reach out to the hypertext community. On 6 August 1991 I posted to the alt.hypertext Usenet group.

> The WorldWideWeb (WWW) project aims to allow links to be made to any information anywhere. . . . We have a prototype hypertext editor for the NeXT, and a browser for line mode terminals which runs on almost anything. . . . If you're interested in using the code, mail me. It's very prototype, but available by anonymous FTP from info.cern.ch.

The response I got from the hypertext community was lukewarm – almost no one had a NeXT, and those who did found the web server software confusing. Still, by the end of the year and after the demonstration at the conference, a handful of scattered websites began to pop up 'in the wild'.

By 1992, with CERN's blessing, I was ready to take web technology to the big time. That year, NeXT, Inc., had organized a developers' conference in Paris, and Steve Jobs was going to attend. With great excitement, I packed the NeXTcube from my office into boxes and drove it to the conference hall in Paris. Alongside dozens of other hopeful developers, I set up my computer on a folding table in the demonstration hall, loaded up my web browser and prepared to take Jobs on a tour of the best websites in existence.

A murmur ran through the hall when Jobs entered. Even then, he was a celebrity. Jobs took his time with each developer, pausing to ask a few questions about the projects, before moving casually on to the next. As he approached my table, I was giddy with anticipation – but before he could reach me, an aide approached and whispered in his ear. Jobs, called away on business, turned and left, never to return.

I was crushed. I'm still disappointed, actually! The World Wide Web succeeded, of course, and so did Steve Jobs, but I never got to meet him. As I grow older, I've come to see this as a kind of 'sliding doors' moment – not just for me, but for the internet as a whole. Maybe I flatter myself, but I believe that if Jobs had just taken a few more steps down the table to see that 1992 World Wide Web product demo, the world might be a completely different place.

Jobs had talked, after all, of 'interpersonal' computing. In fact, that was the whole point of the NeXT. And there, in Paris, I had what would prove to be the great interpersonal computing medium of our time – the NeXT's killer app! If Jobs had seen the early World Wide Web, I believe he would have got it instantly, just as Robert Cailliau and Paul Kunz had. And from there, who knows what might have happened? Perhaps Jobs would have developed the great browser/editor I dreamed of. Perhaps the web would be influenced by Jobs's design philosophy, and evolved into something more stylish, elegant and user-friendly. Perhaps NeXT would be one of the world's largest corporations, instead of Apple. Perhaps I would have even gone to work for him! Perhaps . . . but it didn't turn out that way.

•

Meanwhile, the internet was growing in the USA. In late 1991, the US Congress allocated around $600 million to develop what was then called the 'Information Superhighway'. The principal architect of the funding bill was Al Gore, a senator from Tennessee who soon became vice president. This congressional funding shepherded the early internet from its academic cloister to widespread public use, in particular by encouraging the development of a number of early web browsers. When Gore was later running for president, he rightly took credit for pushing this funding, but his words were misconstrued, leading some to assume that he'd claimed to have *invented* the internet. The remark was widely misreported and ridiculed, and the misinterpretation turned into an urban legend that circulated for years.

Let me set the record straight: Al Gore was critically important to the web's success. He should be honoured, not ridiculed. He had a far-sighted vision for putting technology in public hands, and his funding came at a critical time. He did not invent the internet, but there is no politician, anywhere, who can claim as much influence on the subsequent *development* of the internet as Al Gore. My words echo those of the true inventors of the internet, Vint Cerf and Bob Kahn: 'No one in public life has been more intellectually engaged in helping to create the climate for a thriving Internet than the Vice President.' (Gore and I would eventually meet and share mutual respect.)

While I maintained control of the HTTP, HTML and URL protocols, and a web server which would run anywhere, at CERN we had simple text browsers which only ran on the NeXT computer. We needed browsers which ran on Unix, Macs and PCs. My experience with open-source software suggested to me that if I opened the project, the community would produce a superior product. So I encouraged the development of an ecosystem of browsers, and several developers took up the challenge.

This was the beginning of a long and, indeed, ongoing series of 'browser wars'. The first developer to take the web browser seriously was Pei-Yuan Wei, a computer science student at UC-Berkeley. Pei Wei had independently been thinking along the same lines, looking to use Viola, his own HyperCard-like system, as a platform for networked computing. Then he stumbled across my alt.hypertext Usenet post from August 1991. He recognized the importance of the hyperlink, and in December 1991 he sent me a line, telling me he was thinking of developing a browser for the Unix-based X Window system, a free and open-source graphical interface. 'Sounds like a good idea,' I replied.

The first version of Pei's browser, known as ViolaWWW, launched to the public in March 1992. With my encouragement, Pei continued to develop it, and added several novel features. His browser included 'bookmarks' so users could save important web pages they wanted to revisit. It kept a history of websites you had visited, and it had buttons for going backwards and forwards. It also could run small multimedia 'applets' within the browser, offering interactive content, and had a style-sheet feature which could, among other things, resize web pages to fit the size of the user's screen. These features became standard in practically all subsequent browsers – Pei Wei does not get enough credit for his contributions.

Several other people independently developed early browsers. At the University of Kansas, Lou Montulli debuted the text-only browser Lynx. At CERN, Robert coded MacWWW, the first browser for Macintosh. In Finland, a group of four graduate students at the Helsinki University of Technology published the Erwise browser as their master's project. I actually travelled to Helsinki later to encourage them to continue developing it, but they had no money to fund their work – Finland didn't have Al Gore. I thought Erwise had a lot of promise, and I would have continued developing it myself, but the documentation was written in Finnish, which was decidedly not one of my languages.

As others developed browsers, I pushed for formalization of the HTTP, HTML and hyperlink protocols. In March 1992, I went to San Diego for a meeting of the Internet Engineering Task Force (IETF). This was basically the steering committee for the internet. The IETF regularly produced technical documents outlining how the technology ought to work. (At the time, the IETF was run by the US government, although it became an independent non-profit in 1993.) I was hoping to convince the IETF to publish standards for my new web protocols, encouraging their widespread use.

IETF meetings were organized into both formal working groups and informal 'birds-of-a-feather groups', confederations of affiliated technical researchers who met and hashed out details until they agreed on a particular specification. This collaborative model was not without drawbacks – technical infighting could sometimes delay the publication of important standards by years. I was hoping to avoid such a fate for my protocols, but as the San Diego meeting approached, I was beginning to sense trouble.

There was a push within the working group to rebrand what I had called the Universal Document Identifier as a Uniform Resource

Locator. To me, the terms 'universal' and 'uniform' contained a crucial distinction. 'Universal' meant that anything might appear on the other end of a link. 'Uniform' implied there was some kind of standard that had to be conformed to – which was the exact opposite of what I was trying to achieve. In online correspondence leading up to the San Diego meeting, the discussion grew pointed.

For years, before this event, I had suffered from debilitating back pain. I had slipped a disc in my spine some time before, causing shooting pain to course down my legs. I was taking painkillers for the condition, but in the weeks leading up to the San Diego conference, the pain grew worse, to the point where I could barely walk. In Switzerland, I met with a spinal specialist, who was concerned, and recommended immediate back surgery. At some point I was shown a model of my spine and the spinal cord. I could see the little disc that was impeding the nerve, but I convinced him to delay the operation until after the IETF conference.

The journey from Switzerland to California was a challenge, as I had to keep my back completely straight in the flight seat. In this agitated state, I began to rehearse my arguments for 'Universal' over 'Uniform'. The discussion continued, in my head, up to the day of the working-group discussion. As I got ready at my hotel on the morning of the meeting, I was mentally preparing myself for the battle, when suddenly, in the shower, I felt a pop in my back – the pain was miraculously gone! I attribute this cure to a totally unconscious change in my posture: the 'alpha' stance of a primate ready for combat, standing tall with my shoulders back, attempting to appear imposing.

Unfortunately, after a great deal of arguing, I lost the actual, non-imaginary argument within the working group, and the published IETF standard for web technology defines clickable hyperlinks as Uniform Resource Locators, or URLs, the term by which they were known for ever after. The good news is that I never had to have back surgery, and I stopped taking painkillers. Also, regardless of what the IETF says, *I* still use the term 'Universal'. Just try to stop me.

•

As the early web flourished, CERN graciously relieved me of most of my previous commitments. In a way, CERN was the best place to be. It was located at the centre of Europe, and it was the most advanced physics laboratory in the world. On the other hand, the internet had started in the US, and so the centre of gravity was still there.

I contemplated taking a sabbatical year in the US, but David Williams, the lead of the Computing and Networking division, bargained me down to taking the summer of 1992 as paid leave. There was a sense, for both of us, that the web was outgrowing CERN, so I needed somewhere to land if I decided to make a clean break. I visited MIT in Massachusetts and Xerox PARC in California, evaluating both as potential homes.

MIT was very welcoming, and seemed a good cultural fit. My ambassador was Karen Sollins, a researcher at the school's famous Laboratory for Computer Science. There, I gave one of my first lectures on the web and was met with a warm reception. I also made sure people at the lab were using the web, sometimes even installing the Viola browser on MIT machines myself.

Xerox PARC, in Silicon Valley, was equally impressive. This is where Bob Metcalf had invented the ethernet standard, and where the modern computing paradigm of having a big screen, a mouse and a network connection were first demonstrated with the Xerox Alto personal computer. (Steve Jobs had been inspired to build the Apple Macintosh after seeing a demonstration of the Alto.) We stayed in an apartment on Sandhill Road, famous for all the venture-capital action; SLAC, the physics lab which had been the first to pick up the web outside Europe, was just up the road in Stanford. PARC was built into a mountain in Palo Alto, so you could bike into the top or the bottom of it – I have fond memories of Alice's head bobbing gently from side to side in her big pink helmet as she went to sleep on the back of the bike in front of me. I was unsure, though, about relocating the web to Xerox. While it had a great reputation internationally and a lot of star power, it was pretty corporate. I preferred the more fluid and open atmosphere of the academic lab at MIT.

While we were in the Bay Area, I met with Pei Wei and attempted to convince him to develop ViolaWWW for Windows, where we had no available web browser. If the web was going to be a mainstream technology, we needed to break it out onto the most popular operating system. But our lunch, at a cafe outside San Francisco, was frustrating. Pei Wei saw ViolaWWW as a mere subset of a far more important technology he was pushing, and he didn't care for Microsoft. I respected his perspective, of course – I was pushing an uncompromising, long-shot technology of my own. Still, I pointed out to Pei Wei that people were actually *using* ViolaWWW, so that might be an important area to focus on first. As the conversation continued, I tried and failed to persuade him. The discussion of what to do with ViolaWWW was left unresolved.

On this same visit, I first met Ted Nelson – the one who had coined the term 'hypertext'. Ted was an eccentric, wide-eyed dreamer who lived on a houseboat in Sausalito, just north of the Golden Gate bridge. His 1983 book, *Literary Machines*, was an eclectic, non-sequential rant with some very specific ideas about the systems we should be using. Ted had sent me a physical copy at CERN, and in return I sent a physical $25 cheque – but the American banks had no concept of a Swiss cheque, so Ted was unable to cash it. Owing Ted $25 was my excuse to meet him.

One of my favourite ideas of Ted's was that of the two Gods of (Electronic) Literature: the Reader and the Writer. They were both omnipotent in their own way. The Writer could write whatever they wanted to, without constraint. The Reader could read whatever they wanted to, or not read anything. Literature is a battle between these two.

Ted's proposed Xanadu technology would create a new relationship between the Reader and the Writer. It included a property he called 'transclusion': every time you quoted one document in another document, there should be a bidirectional link between the original and the quotation, allowing you to switch between the two. Xanadu also included 'micropayments' (a term Ted coined) to support artists and creators. Click on a link in Xanadu, and the person (or persons) who'd designed the page you visited would receive a small amount of money you could explicitly pay. Ted's dream was that all the literature in the world would be contained in a centralized system run by the Xanadu Operating Company.

Both Xanadu and Ted had a significant fanbase. The project was developed by an informal group of enthusiasts for a while, then adopted by the company Autodesk. On the very day I first met Ted, Autodesk announced it was pulling the plug, and spinning it out again to the informal development community. This was an unfortunate coincidence: a sad day for Ted, and a philosophical one. But I was certainly happy to meet him face-to-face, having heard so much about him.

Ted was a cordial host who put a brave face on the unwelcome news. He and I ate lunch together and afterwards, in the parking lot of an Indian restaurant, I asked whether I could take photos of us with my camera. He replied yes, absolutely – he had been wanting to ask if he in fact could make a video. Ted, years ahead of his time, took videos of many of his conversations. He would begin each video by pointing the camcorder at himself and describing the date and identifying his interlocutor, then swing the camcorder towards his guest to record his conversation. In this way, the appearance of his own head formed an index of the conversations on his videocassette. So I took pictures, he videotaped me while I described the technology behind the web, and we parted.

•

The web's biggest competitor at this time was a system called Gopher, a competing network protocol that did some of the same things, although it used a menu-driven navigation system rather than hyperlinks. Gopher was developed in 1991 by Mark McCahill and Farhad Anklesaria at the University of Minnesota. (The school's mascot was a buck-toothed gopher named 'Goldie'.) Gopher was developed as a campus-wide information system and had succeeded by making available trees of information for administrators. The protocol was similar to HTTP in that it was fast and light; it proved immediately popular on campus and by 1992 had begun to percolate to the wider computing community. In fact, up until 1994, more people used Gopher than the World Wide Web.

Gopher's popularity was driven by the Legal Information Institute, a free database of US Supreme Court decisions. Developed by Tom Bruce, a software engineer at Cornell University, the LII was to Gopher what the SLAC archive was for the web. I spent a lot of time thinking – well, worrying – about Gopher in 1992. It seemed to me that hyperlinks were a much better way to navigate the internet than drop-down menus, but there did not seem to be an obvious way for my technology to catch up.

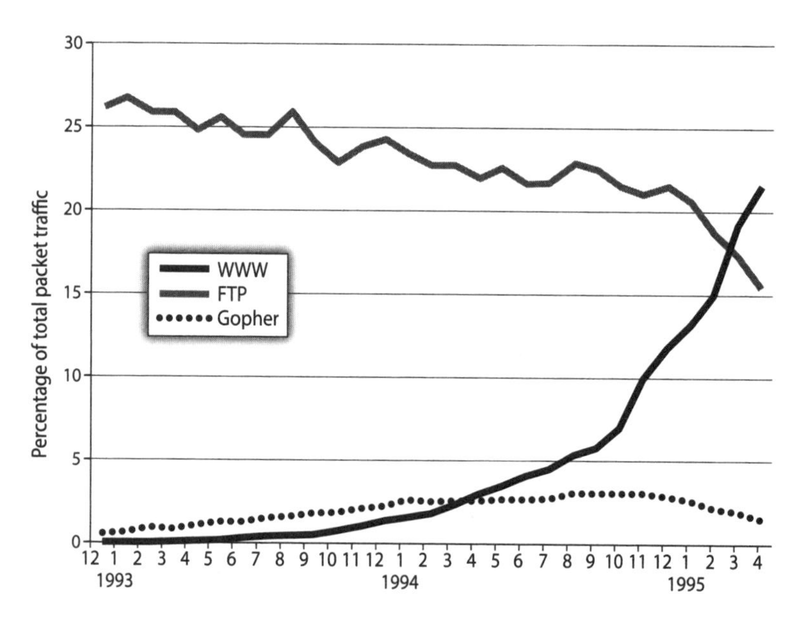

Traffic on the NSF backbone by protocol type, showing the web transitioning from the lowest to the highest traffic.

Then, in February 1993, the University of Minnesota made a fatal mistake: they announced that they were reserving the right to charge licensees to use Gopher. The backlash from the user community was swift and harsh. The predominant ethos of the time was that the internet should be free, and this ethos was enforced with an ideological fervour that might seem mystifying to modern readers. As far as I know, the University of Minnesota never actually collected any licensing revenue from Gopher – they merely said they *maybe* would do so in the future. But that was enough to kill the technology.

Gopher usage saw a drop following the announcement, while web usage saw a spike. To add insult to injury, many people downloaded the tools to access the web using Gopher technology before porting over. Tom Bruce ported his LII database of Supreme Court decisions to the web and began developing the first-ever Windows browser. Riffing on Viola, he called it Cello.

But all this activity only made me more concerned – because the web was in the same position as Gopher! To date, all of the internet protocols had been free and all of the code was open-source. My assumption had been that it would always remain that way – but now, I realized, I couldn't provide this reassurance to the web community. I had developed the web technology without giving much thought as to who actually owned it. As it took off, I understood that the answer was: probably not me. Sure it was my idea, and I had written the code for the protocols – but I had done so on CERN-owned computers, while collecting a CERN salary. The reality was that the web belonged to CERN, and that if someone above my pay grade at that institution had a mind to, they could probably have started charging to use it.

The only way forward, I realized, was to donate the intellectual property behind the web to the public domain. I had never intended to profit personally from the technical protocols I'd written for the web, so this decision didn't impact me – and it didn't matter. For the web to succeed, it *had* to be free.

Vint Cerf had similarly licensed his TCP/IP protocol for eternal, free public domain use, with no restrictions. Cerf, whom I greatly admired, hailed from the mainframe era of computing. Raised in Connecticut and educated at Stanford, he had started his career working for the ultra-square IBM. As the rest of the industry relaxed into jeans, T-shirts and sandals, Cerf, with a certain sense of irony, upgraded to a three-piece suit and pocket square, which he wore for the rest of his career. But Cerf's conservative mode of dress disguised a radical vision. Even in the 1980s, he believed the internet would one day be everywhere – on everything, on every electronic device. At a memorable keynote presentation to the IETF, he'd once performed a striptease, removing his jacket and waistcoat to reveal a T-shirt that read 'IP ON EVERYTHING'. This seductive prophecy has today come true. In the modern world, billions of devices – refrigerators, toys, speakers, printers, watches and televisions – all have IP addresses.

What Cerf was for the internet, I realized I could be for the web – not its chief executive, but its chief evangelist. The web on everything, everything on the web! Robert Cailliau and I began to petition CERN to release the source code for the web into the public domain. This was not an easy task, mostly because it was not clear who at CERN had the authority to sign off on such a decision. Eventually, it was decided that Walter Hoogland, the particle physicist who served as CERN's director of research, would have the final say.

Now, there were those within CERN who felt I was making a mistake. In a world dominated by Silicon Valley, this group of patriots believed that the web should be spun off as a company, to prove that Europe could compete. I was sympathetic to this point of view, but I had seen what had happened to Gopher, where, I suspect, someone at the University of Minnesota had made a similar argument. In search of these phantom riches, Gopher had been driven into the ground.

More importantly, there was the public good to consider. The internet had so much potential to bring people together. Peering into the future, I could see a day where everyone was coming online. Should they be segmented into for-profit, walled gardens? Should they be bound by conventional barriers of nationality and language? Was that what the internet was for? Not if I could help it.

After a fair amount of wrangling, Hoogland agreed that CERN should give up its rights to the web. On 30 April 1993, CERN published the complete source code to my protocols, as well as my client and my web server software, accompanied by the following statement, signed by Hoogland and Helmut Weber, CERN's director of administration:

> CERN relinquishes all intellectual property rights to this code, both source and binary and permission is given to anyone to use, duplicate, modify and distribute it.

Why did I do this? My motivation went beyond avoiding the fate of Gopher. The internet was not so commercialized in those days; in fact, it wasn't commercialized at all. I was still at a research institution, and I embraced the academic culture of keeping things accessible and open. To patent the technology seemed like an affront to the great many people who had worked tirelessly, and often anonymously, to make the internet work.

There was also the question of adoption. It was already a big ask to request that everyone name every document with 'http'. To ask anything else – to charge – was too much. I wanted everyone to use it. I wanted it to be universal. I saw the technology and I saw how effective and easy to use it could be. Above all, I think, I just had a vision for it, a vision of all existing systems talking to one another, of everyone coming online.

Sometime late in 1992, I stumbled upon a wonderful website that contained an archive of illuminated sheet music from the Italian Renaissance. I remember clicking on one thumbnail and having it suddenly appear as a beautiful, full-colour scan on my 21-inch screen. I was delighted – I felt as if I were sitting in the Vatican archives, paging through this antique document myself. The project was a collaboration between the Vatican, the Library of Congress – which had had a physical exhibition and scanned the pictures – and a Dutchman who knew about the web. Not the first time people would collaborate across the internet, but a dramatically beautiful example.

•

In January 1993, Marc Andreessen, an undergraduate student at the University of Illinois, sent a note to the www-talk mailing list, soliciting test users for a new browser he called 'Mosaic'. Mosaic was a Unix-based browser that started simply but evolved rapidly, with updated versions delivered every couple of weeks. I added a download link for Mosaic to the info.cern.ch server and encouraged others to experiment with it.

In February 1993, Marc proposed adding the IMG tag for inline HTML images. In the past, browsers had rendered images in a separate window, but Marc was proposing to include them on the page and render them on the fly. I liked this idea, but wanted to do it in a more general way (reusing the link for *any* media item). My objection to the IMG tag was based on my preference for clean powerful design, rather than an objection to images per se. A spirited discussion of the merits of our proposals followed – at one point a Dutch developer named Guido van Rossum chimed in. Marc later became one of Silicon Valley's leading venture capitalists, and van Rossum invented Python, today the leading general-purpose programming language for artificial intelligence.

Marc eventually just unilaterally added the IMG specification to Mosaic. As was his right, I suppose – the whole purpose of placing the web standards in the public domain was to allow for this sort of modification, without my explicit permission. And Mosaic was fast becoming popular – a dominant monopoly.

Mosaic rapidly surpassed ViolaWWW as the web's most popular browser. A great part of its success was due to Marc's hard work. He paid close attention to user requests from various hypertext newsgroups and, together with programmer Eric Bina, rolled out features on a rapid development cycle. Marc was the first to approach web users as 'customers' and, unlike Pei Wei and Tom Bruce, he had a team of people supporting him – funded by Al Gore's bill. Mosaic also included a regularly updated landing page entitled 'What's New' which provided a chronologically organized list of new web servers. By the middle of 1993, new pages were appearing every day.

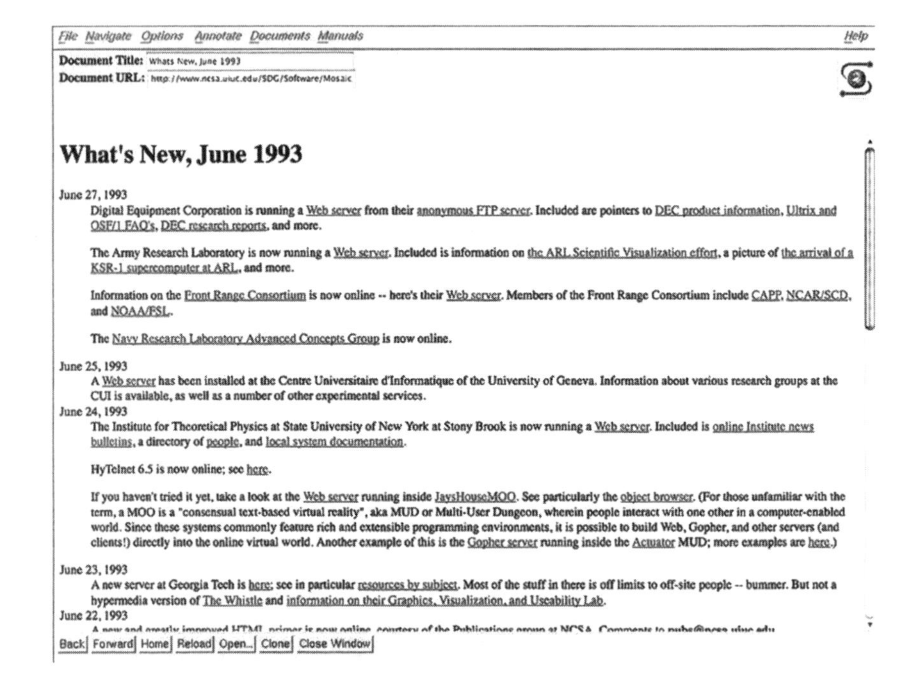

A screenshot of the 'What's New' page in 1993, listing recent additions to the web.

When I first saw Mosaic I was thrilled, but as the software became more popular, I began to grow concerned. In talks at conferences, the Mosaic team seemed to be deliberately avoiding any mention of the World Wide Web. Rather than referring to pages being 'on the web', they referred to them as being 'on Mosaic', burying all mention of the protocols I'd conceived. In March 1993, I was invited to the Fermilab particle accelerator outside Chicago to deliver a presentation on web technology. By coincidence, Tom Bruce, who hosted the first online legal archive, was in Chicago at the same time, and he showed me an early version of his Cello browser. Following the presentation, Tom and I, together with Ruth Pordes, my host from Fermilab, decided to drive from Chicago to the National Center for Supercomputing Applications (NCSA) in Champaign, Illinois, where Marc and his team worked. The drive took us through several hundred miles of cornfields; having lived surrounded by mountains for almost a decade at this point, I was struck by how flat it all was.

The meeting took place in the basement of a petroleum chemistry building. Tom, Ruth and I sat opposite Marc, Eric and Joseph Hardin, the leader of the NCSA's software development group. All of my previous meetings with web developers had been cordial, collaborative and enthusiastic, but at this meeting, tension hung in the air. Picture us together in this dreary, windowless room: Marc was very young, and he still had all his hair. (I retained some of mine as well.) In the days before the meeting, it had become clear to me that the NCSA group was attempting to hijack my creation, and rebrand the web as Mosaic. Politely, I articulated my displeasure at this effort. Marc grew visibly uncomfortable, but said very little. He seemed to regard the meeting as some kind of poker game.

I dropped the subject, and attempting to foster a spirit of collaboration, I asked Marc whether he had considered adding an editor feature to the Mosaic browser. Marc and Eric responded that this was 'impossible' (it wasn't, I had done it). He also argued that, in any event, users were demanding more multimedia features, not editing capabilities (this was true). As Marc, Eric and I continued this discussion, I began to realize there was a second tension in the room – their supervisor, Joseph Hardin, kept interrupting them. Hardin had previously developed a multimedia hypertext editor called Collage, and he seemed to regard Mosaic as a sequel to work he'd already done. Soon, it became apparent that Marc was just as concerned about Joseph taking credit for *his* work as I was about Marc taking credit for mine. In general, he did not seem to appreciate Joseph as his boss.

I left this meeting dispirited. My concerns were not assuaged the second time I met Marc. In summer 1993 I travelled to Cambridge, Massachusetts, to attend the Wizard's Workshop, the first web meet-up. The workshop was organized by Dale Dougherty, who worked for the tech publisher O'Reilly, who was using it as a platform to launch the first commercial web publication, the Global News Network. (GNN is today best remembered for featuring the first clickable advertisement.) Marc was in attendance, and while the other Wizards were collegial, he seemed to be keeping his distance. Several attendees remarked how different his combative online persona was from his personality IRL – in real life.

Soon it became clear that the NCSA had big ambitions. Shortly after our second meeting, Mosaic was ported to Windows, triggering an avalanche of public interest. By late 1993, Mosaic held an insurmountable lead in the browser sweepstakes, and the NCSA organized a public-relations push around it. In December, *The New York Times* ran its first mention of the web, under the headline 'A Free and Simple Computer Link'. The article described what you could do on the web in 1993:

> Click the mouse: there's a NASA weather movie taken from a satellite high over the Pacific Ocean. A few more clicks, and one is reading a speech by President Clinton, as digitally stored at the University of Missouri. Click-click: a sampler of digital music recordings as compiled by MTV. Click again, et voila: a small digital snapshot reveals whether a certain coffee pot in a computer science laboratory at Cambridge University in England is empty or full.

This text ran below a picture of Joseph Hardin and NCSA director Larry Smarr. Mosaic got most of the focus. My name wasn't mentioned until well into the article; Marc and Eric weren't mentioned at all. Still, the article was a big, big thing for the web – it showed we were gaining momentum. Shortly before the article appeared, I had attended the Hypertext '93 conference, hosted in Seattle. No need to jerry-rig an internet connection this time! Every single project on display there had something to do with the web.

CHAPTER 5

Governing the Web

At the beginning of 1993, there were about fifty web servers in the world. By the end of that year there were more than 600, and the original server at info.cern.ch was getting 10,000 page views a day. Statistics from the National Science Foundation showed that in March 1993, web protocols accounted for about 0.1 per cent of all internet traffic. By the end of the year that had risen to 2.5 per cent. No other protocol had ever grown so quickly.

Vint Cerf, the co-inventor of the original internet protocol layer, became aware of the web protocol around this time. Cerf's previous efforts to put multimedia objects on the internet had proven expensive and unwieldy. Most hosting formats were proprietary, and you often had to buy extra software to interact with the object. 'The web created a layer of loam, if we want to use an agricultural analogy,' Cerf recently recalled. 'The beauty of it was the browser – you could see what other people had done. The webmasters learned from each other by inspecting HTML. It's one of the best examples of why open-source is a powerful tool.'

As the web grew, the number of websites in existence far exceeded my ability to monitor them. By design, the web had no central registry, although in the early days most people who hosted a web server took the courtesy of informing me via email. By 1994, though, websites were popping up all over. In January that year, Jerry Yang and David Filo, two graduate students at Stanford, started a directory of web pages organized by topic, then broken down into subtopic. They named the directory 'Yet Another Hierarchically Organized Oracle', or Yahoo. At first, there was no search function – the two added new sites to the Yahoo directory by hand.

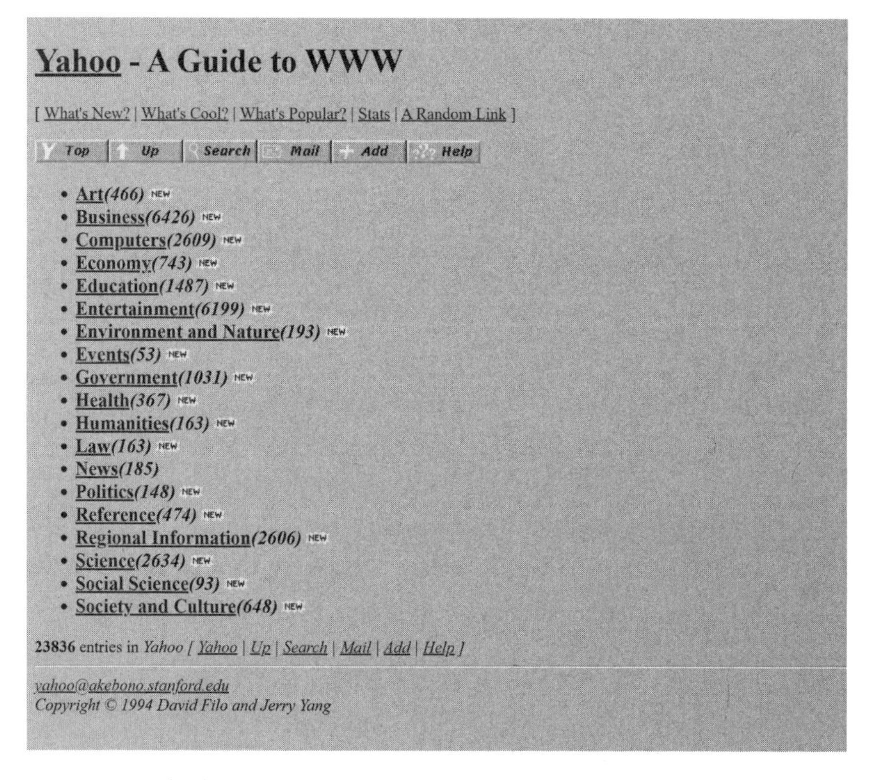

The home page of the Yahoo search engine in 1994.

The rapid growth of the web drove the exploding public interest in the internet. In addition to Yahoo, retail internet service providers like CompuServe, America Online (AOL) and Prodigy were beginning to appear in the US, offering users paid monthly access via dial-up modems. Originally, many of these services were 'walled gardens' that sequestered customers into digital channels that the ISPs controlled, but users began to demand access to the general internet. In early 1994, AOL opened access to Usenet newsgroups like alt.hypertext.

Up to this point, the internet had mostly followed the norms of academia, resulting in a measured (although occasionally rather sharp) standard of discourse we called 'netiquette'. Each September, a freshman class would arrive on campuses and be granted internet access, and there would be a rocky adjustment period as the new arrivals adjusted to these informal rules. The arrival of the public, via AOL in 1994, destroyed netiquette for ever, replacing it with something far more raucous, funny and crude. Seasoned internet users called this change the 'Eternal September'.

While some computer scientists objected to Eternal September, it was fantastic for the web. I had deliberately designed my protocols and creative tools to be simple to use, and a great many new sites appeared, often coded by hand in HTML, sometimes by users who'd only just learned how to use a computer. These delightful amateur sites weren't bound by academic conventions and often showed remarkable creativity. In January 1994, Justin Hall, a freshman at Swarthmore College in Pennsylvania, began using the web to document his personal life, in what was perhaps the first 'blog' with comments. He called his site Justin's Links from the

Underground, and in addition to his musings, he also posted multimedia objects he'd discovered, including bootleg recordings of his favourite band, Jane's Addiction, a picture of Cary Grant dropping acid, and a dedication to Al Gore: 'The Information tollroad's first official pedestrian'.

Will it surprise you to learn that this is exactly what I had intended the web to be? Well, not *exactly*, of course – I could never have predicted Justin's creativity or his sense of humour. But I had hoped to usher in a new era of creativity and collaboration, and early publications like Justin's were only the beginning of the astonishing activity that was to come. It was all about unleashing *other people's* creativity and imagination: if the web only produced things I could imagine myself, it would have failed.

•

My fellow evangelist Robert Cailliau had long pushed for CERN to host a conference for web developers, and in late 1993 the institution agreed. Robert publicly announced our conference shortly after Hypertext '93. But then, the next day, Hardin and the NCSA announced that they would be hosting a conference, too. I perceived this as an overt attempt to hijack my creation. As the kids might say – it was *on*.

I felt it was critically important that *our* conference happen first, and in Switzerland rather than Illinois. The earliest CERN could host the event was May. Robert booked the auditorium and meeting rooms, and emailed potential invitees – including Joseph Hardin at the NCSA. Hardin quickly responded that *his* conference was also in May, and that he was inviting many of the same speakers.

Following a rather tense standoff, we agreed that the first International Conference for the World Wide Web (WWW1) would be held at CERN in May 1994, and that the second, WWW2, would be held in Chicago in November the same year.

So we won the conference showdown. To me, this was a crucial moment in asserting my vision for the web: not that I should control it, but that *no one* should control it. I could already see that Mosaic wanted to dominate the browser space and unilaterally dictate new HTML standards. My fear was that this would lead to a splintered web with multiple communication protocols evolving independently. In the worst-case scenario, some browsers wouldn't be compatible with some websites – a real Tower of Babel situation. To avoid this, the web needed some kind of governing body.

As the conference date approached, I began to reach out to various institutions about the possibility of hosting a web *consortium*, a coalition of stakeholders who oversaw the ongoing modernization of the HTTP, HTML and URL protocols. Nobody would own these protocols; the consortium would host working-group discussions and regularly publish technical specifications delineating best practices for new web technologies. Most importantly, the standards would be royalty-free. I called this, at that point, the World Wide Web Organization, or W3O, although it later became the World Wide Web Consortium, or W3C, as it is known today.

Now, this may seem a roundabout way of doing things. Why not form a company? Why not just dictate the standards myself? The answer was that those approaches inevitably led to Balkanization and competing technical fiefdoms, and that was the exact opposite of the *universal* nature of the web. If everyone was going to use the web, then everyone had to have a say in what the web was. I had learned a little from the IETF meetings I had attended, and MIT had started a number of consortiums too, so I had some sense of how it should operate.

There was also the looming question of multimedia. A state-of-the-art home computer in 1994 would have a VGA monitor with 256 colours, a sound card, and a 28.8 kbps (kilobits per second) modem. With this expensive technology, the home user might – *might* – just be able to render Justin Hall's Links from the Underground, assuming they were patient enough for the images and audio elements to download. The early web demonstrated more of a yearning for multimedia than actual content, but that yearning was so strong, and that potential so powerful, that I could already feel the pressure to update the HTML protocol continuously, as rapidly as possible. W3C would give us a framework for doing so.

But I had to move quickly – already, I could see that the NCSA wanted to set the standards unilaterally themselves. Allowed to do so, I expected they would attempt to profit from it. That could lead to fragmentation, with my 'free' set of web standards competing with the NCSA's 'expensive' one. This absence of interoperability had, throughout the history of computers, led to all sorts of annoyances for users. Older readers will remember the incompatible file systems of PC and Macintosh, and all the irritations that were created. A more modern example would be Apple's blue text message bubbles versus Android's green ones; if you've ever been in a texting thread with users of both platforms, you'll understand the headache. Only a consortium of stakeholders – with everyone at the table having a voice – could avoid this profit-driven splintering. There had to be just one web.

The question was where to host this consortium. One possibility was CERN, but by 1994 I felt the web was outgrowing my institution. CERN had been a wonderful place, and had offered me time, resources, attention, and no small amount of personal encouragement and warmth. But its core mission was to explore the limits of physics, not to manage internet communications protocols. To host the web there, I felt, would be a distraction.

If it wasn't CERN, then it had to be somewhere in America – my server logs clearly showed that that was where the traffic was coming from. Silicon Valley was an option, of course, but as the web was intended as a not-for-profit standard, it didn't seem like the best cultural fit. Around this time I attended a conference in England, in Nottingham. It was pouring with rain, and the attendees had to be transported around in buses. By coincidence, on one of those bus trips I was seated next to David Gifford, a professor at MIT. I talked about the choices I was making about the future of the web, and about the need for some kind of governing organization that would nurture the web and keep it unified – and about locations and hosts. He recommended I talk to Michael Dertouzos, the head of the Laboratory for Computer Science at MIT. That, of course, was the same MIT lab I had visited two years before.

Michael was Greek, and he visited Europe often. In February 1994 we discussed our ideas for how a web organization might function over my favourite dish of Zürcher Geschnetzeltes Kalbfleisch and rösti (the classic Swiss dish of sliced veal with a cream sauce) at a cafe in Zurich. (Michael, a large man, was a hearty eater.) MIT had successfully run such an organization before, for the widely used Unix X Windows system, and this provided a template for what I wanted to do.*

A short time later I received a surprise visit from Alan Kotok, along with several of his colleagues. Alan, a consummate geek, worked for the Digital Equipment Corporation, but in his spare time he was a big mover and shaker in the MIT Model Train Society. While this was notionally a DEC sales call, I sensed it had an ulterior purpose – a suspicion confirmed when Kotok started praising my W3C idea and suggesting what a great home MIT would be for it. The charm offensive had begun.

MIT was attractive in other ways as well. Alice by this time was in pre-school, and I had observed how frequent the turnover was among her classmates. Geneva was home to a large number of expatriate families, and the children in these families often moved several times before adulthood, as their parents bounced around between international agencies. My French teacher called such families *fleurs sans racines* – flowers without roots. There was also my wife Nancy to consider. She had grown up in New England, and was now pregnant with Ben. Lately, she'd expressed a strong desire to return home.

•

The First International Conference on the World Wide Web kicked off on 25 May 1994. The atmosphere was electric. In the months before, we had put out a call for conference attendees and were deluged with

* There is some irony there, in that the X Windows system was having a rough time, as Windows and Mac seemed to be eclipsing Unix in general. Bob Scheifler, the creator of X Windows, closed the X Consortium just as Linux took off.)

requests. Perhaps twenty-five people had attended O'Reilly's Web Wizards Workshop. More than 800 people wanted to attend WWW1, but CERN's space held less than half of that, so with great regret we had to turn a lot of people away. In the end, about 350 people attended, including members of the press and gatecrashers. Many people who'd spent years corresponding online were able to meet one another for the first time; some later called it the 'Woodstock of the web'. I thought that a little over the top, although I certainly had a good time.

The attendees drew mostly from the hypertext community and 'veterans' of the web – loosely defined as anyone who'd been using it for longer than a year. (There were also some members of the business community in attendance, but fewer than you'd think.) Dave Raggett, a programmer with Hewlett Packard, showed off his Arena browser, which brought a 'magazine-like' sensibility to web pages, with support for background images, tables, and text flow around captions. Meanwhile, in breakout sessions in the meeting rooms, technical experts gathered to discuss the future of HTML. Over the course of a few hours, we developed a roadmap for its development, including support for many of Raggett's innovations. Another session discussed technical issues related to HTTP, which was rapidly becoming the single-most popular internet protocol. The collegial nature of these discussions, and the rapid, collaborative development of product blueprints that met the needs of a great variety of users, gave me confidence that the consortium was a good idea.

Brandon Plewe, a student at Brigham Young University in Utah, had, in the lead-up to the event, proposed several categories for 'Best of the Web' prizes, and solicited votes via email. Having underestimated the web's popularity, Brandon received 5,225 ballots, which he'd had to tabulate by himself. The prizes were awarded at a dinner on 26 May. One prize went to Steve Putz at Xerox PARC, who'd built a brilliant interactive 'map viewer' that generated geographical mapping services on demand. Another went to Nicolas Pioch, a French student who'd built a virtual copy of the Louvre, complete with high-resolution digitizations of the Mona Lisa and *The Death of Marat*. Another went to Oliver McBryan at the University of Colorado, who'd coded the World Wide Web Worm, one of the first search engines. Although these ideas were in germinal form, each would later become a ubiquitous technology.

The dinner also featured the inaugural induction ceremony to the World Wide Web Hall of Fame. Six of us were inaugurated, including Marc Andreessen and myself. The following day featured a panel discussion on the future of the web – standing room only, as you can see.

The first web conference at CERN. The main auditorium was packed!

As the web's inventor, I was asked to deliver the wrap-up speech. In the weeks leading up to the conference, I had begun to consider some of the web's emergent properties, properties which only became apparent at scale. The web was growing at an exponential rate – the classic hockey-stick graph. To really see what was happening, you had to plot the web on a log scale chart, with each interval on the y-axis representing 10x growth. Only when you did this could you see adoption moving in a straight line.

The thousandth person who joined the web was doing so for a different reason from the hundred thousandth person, and the billionth person who joined the web would do so for a different reason from the hundred thousandth – but as the web grew, I was coming to understand that each individual web user was only a part of the picture. Equally important were the unexpected and organic *links* that brought the entire network together. In my speech I outlined these properties:

> The dictionary defines words only in terms of other words. What matters is not the words themselves which are a collection of mostly arbitrary sounds, but the *relationships* between one word to the next. It is a common view, in computer science, to define the world in terms of information, but a single piece of information is meaningless. . . . In an extreme view, the world can be seen only as *connections* between information. This resembles the neurons in the brain, but the brain has no knowledge until connections are made *between* the neurons.

During this speech at CERN, I also talked about the differences between documents and data. People understand documents; computers understand data. Documents can contain all kinds of information, often described in natural language, and a person can understand how those items relate to one another. But unless those documents are expressed in terms of data, a computer cannot do the same. Take a bank statement. On the one hand, you can download it as a spreadsheet from your bank. Your computer, via Microsoft Excel or Google Sheets, immediately understands what it is and how to read it. It just contains hard data. However, if you've just got a paper statement from your bank, or even a scan of it, you have a document. Your computer doesn't see the data within – at least without further processing.

By 1994, it was clear to me that we were moving away from just a web of documents towards a web that included *data*: well-defined relationships between well-defined things. I called this the *semantic* web. 'Semantics' means the study of meaning. What I could see was that, within all those documents online there was an extraordinary amount of data that captured relationships between people, objects, places and ideas. If we could somehow structure that data, we could then make it readable by computers. And from that step, I figured, we could make computers *reason* about these relationships in an intelligent way. It was a bold vision – in some ways bolder than the World Wide Web itself. The concept of the semantic web, and the ups and downs it experienced throughout the following decades, will crop up many times in this book.

•

Around the time of the web conference, I'd been fielding a lot of calls from Al Vezza of MIT. Al, who had previously been the CEO of Infocom, publisher of the legendary computer game *Zork*, worked under Michael Dertouzos at MIT. He was the point person for actually establishing W3C. Setting it up was easy; all Al had to do was take the membership contract for the X Consortium and change every instance of 'X' to 'WWW'.

At the same time, I hoped to open a second office of the consortium somewhere in Europe – it was important to me that the web be a global institution, rather than one controlled exclusively by Americans, as so often happens in tech. To my great surprise, Michael not only agreed to this idea, he absolutely loved it. (I suspect he was influenced by his dual citizenship of the US and Greece.) So even as I sought space at MIT, I simultaneously was developing a second office for W3C at the National Institute for Research in Digital Science and Technology (INRIA) in Rocquencourt and Sophia Antipolis, France. The French computer scientist Jean-François Abramatic managed that side of things; he became a good friend and later the W3C chair.

A month after the WWW1 conference, Ben was born, and Nancy and I began preparing for our annual vacation to New Hampshire, so that her parents could see the new baby. In early July, Al called me in a state of excitement – he'd got the go-ahead from MIT. The institute was prepared to hire me as a full-time staff member and allocate space and funding for W3C. The holiday turned into a transatlantic move. We took a dozen or more boxes and suitcases, with a caravan of friends' cars to get us to the airport. The rest of our belongings could follow in due course. We settled on a start date of 1 September.

Al had called again on Bastille Day, 14 July, to go over more details. I half-listened to him as the fireworks went off around me. This was one of those calls that bridges totally different things, totally different cultures, totally different places, the past and the future. I was saying goodbye to CERN, goodbye to Europe. In the distance from my house, across a grazing field for livestock, with the fireworks behind me, one could make out the darkened outline of Mont Blanc. I would maintain a foothold in France, through INRIA, but the centre of mass of the web was shifting towards the other side of the Atlantic.

CHAPTER 6

The Wave

Around this time I got a little chubby. I was flying around a lot, often jet-lagged, had a new baby, was working constantly, and eating too much nice food. My windsurfing hobby didn't follow me to the North Atlantic coast. Boston also offered fewer mountains than Geneva. However, it was a great place for fitness, so in 1995, as I approached the age of forty, I took up running.

Now there are two types of runners: cats and dogs. Cats have set running routes with little deviation, and they focus on performance and consistency. Dogs run off into the woods without a plan, in search of whatever they find interesting at that moment. It will not surprise you to learn that I am a dog; rarely do I return from a run without my shoes covered in mud.

Running provides me with a way to keep fit, but it also gives me a swift way to cover new territory. When I arrive in a new city, one of the first things I will always do is go for a run. I almost never have a set route; instead, I follow what I perceive to be the natural flow of pedestrian traffic. Typically, this will bring me to a river, as most cities, historically, organized themselves around rivers, which are natural engines of industry and trade. The best way to understand a new city quickly is to understand its waterways. I've run along the Thames, the Elbe, the Spree, the Guadalquivir and the Seine. In Boston, I got to know the city via the Charles, which runs past MIT and the West End to the harbour. The landmarks remain fixed but the rivers themselves are turbulent networks of change. As Heraclitus observed, you can never step in the same river twice.

•

One of my goals with W3C was to bring together a broad variety of perspectives, to recreate the 'Medici effect' I'd benefited from at CERN. The objective was to have corporations, small businesses, non-profits, educational institutions and governments, all at the same table. We offered two membership tiers – $50,000 a year for large corporate members, and $5,000 for everyone else. Within the consortium, both tiers were treated the same.

This organizational design was very purposeful. Obviously, W3C needed money to function, but I felt it vitally important that this objective take a firm second place to humanistic values. By the mid-1990s, it was obvious to some that fortunes were going to be made on the web; what was perhaps more obvious to me than to others was that the profit motive and possible monopolies might corrupt what I was trying to build. My feeling was that the most meagre non-profit should have as much of a platform as the largest corporation. It was just so vital that the technology directed users towards humanistic, rather than material, concerns.

The management structure of W3C was initially simple. I was

the director, a lifetime appointment. Al Vezza was the first chairman. I drafted the technical agenda, and he recruited new members. The members would, it was hoped, through online discussion and annual in-person conferences, reach a rough consensus about emerging technical standards. W3C would then publish working drafts for these standards, which would go through a few rounds of editing, before becoming the official standard that we would endorse.

There were some within the consortium who thought this method was too slow for the rapidly evolving technology of the web, but from my perspective, it was the only way to avoid fragmentation. And it worked – even today, everyone on the web uses the same HTTP protocols and HTML standards that W3C endorses. Even some of the very powerful players that would later emerge – Apple, Microsoft, Google – still adhere to these standards of interoperability. It's remarkable to me that in a world driven by profit, both the web and the internet still survive via this collaborative model.

I achieved success at W3C by listening. I wanted the consortium to function, and my belief was that these institutions could find, among themselves, the ideal mutual solutions. Now there were a great many technical arguments along the way, similar to the arguments with my opponents at the IETF that I'd rehearsed in the shower in San Diego. As director, I reserved the final say – if there was an argument that absolutely could not be resolved through working-group discussion, I had the power to make a unilateral decision. But in the nearly three decades I served as director of W3C, I only used this power a couple of times.

The consortium operated through regular meetings, first at MIT, and later throughout the world. Typically hundreds of people would attend, beginning with an all-hands symposium then proceeding into working-group discussions. I soon recognized that we would need a third W3C host to manage the burgeoning interest in the web from Asia. Since 1996, Keio University in Tokyo, Japan, has been our gracious host. And in 2013, Beihang University in Beijing, China, became the fourth W3C host. By establishing ourselves in academic settings, we were best positioned to provide a neutral ground for competing interests.

We soon had more than 100 paying participants in the consortium, which provided enough revenue to hire a small but extremely talented staff. We also received early financial support from DARPA – the US Department of Defense's research arm, which also funded research at MIT – and the European Commission.

Always in the background – and sometimes in the foreground – was Michael Dertouzos, the head of MIT's Laboratory for Computer Science. Michael was a great mentor for me. With his loud voice and expressive Greek personality, he would always dominate the room. I thought of him almost like an admiral – a commander. A natural salesman, he was always pushing for greater attention and resources for the MIT lab. Researchers at MIT sometimes resented his salesmanship, but you had to admit he was great at fundraising. He used to wear a T-shirt that read 'World's Greatest Guano Harvester', a reference for his talent for bringing in the money.

Money to Michael was a necessary evil. He lived very big, and would always park his new BMW outside the lab. When he drove down to the Hilton, he would leave his car right outside the hotel and leave a giant tip for the valet. He liked fine dining; he liked to fly first class. I didn't adopt these habits, but I did learn from Michael the importance of earning one's keep. MIT salaries were paltry compared to what most of us could have earned in industry, so he pushed all of us researchers to supplement our income with money from the lecture circuit. As the years progressed, he even organized speaking tours for the MIT faculty, knowing that otherwise academia would never pay enough to keep them.

Michael taught us not to use slides in our presentations. 'Never use slides,' he said. 'It is you they want to see. If you feel you need your slides as a reminder, here is an experiment. Just imagine slide three; if I gave you the title of the slide, you know you'd be able to talk about it. So just go in with a list of the titles of the slides.' The next time I gave a talk, in Paris, I just wrote down the titles of ten of my slides on the little Psion Personal Digital Assistant I carried everywhere. The talk was a success. 'Ten lines, that's all, that's how to do it,' I thought to myself.

Once the web took off, I began to receive hundreds of requests for paid lectures a year. I decided I would do no more than twelve, annually. Michael, in addition to being a mentor, became a protector for me as well, helping me transition to a more global sort of life. I began to travel with him frequently, and we always had entertaining conversations on the plane.

•

Early on, we had to select an internal computing platform for W3C. Following poor sales, Steve Jobs's NeXT hardware had been discontinued in 1993. The company had then licensed its NeXT operating system for use with generic computing equipment, so, not wanting to switch environments, I ran NextOS on a Hewlett Packard workstation. Later we switched to Microsoft Exchange Server, so that we could be using the same system as the majority of the world. People were encouraged officially to use Windows, but the system team wasn't impressed. The joke was if you left your laptop with them for a moment it would end up being converted to Unix. The final straw for me was when the Exchange tools kept just hanging and not letting me know what was going on (like a web browser would).

In the end, there was a rebellion, and we all switched to using Unix – and the web.

One early hire at W3C was Hakon Lie, a Norwegian programmer I'd previously worked with at CERN. Hakon was a quirky, delightful guy who kept an inflatable reproduction of the figure from Edvard Munch's *The Scream* in his office. Hakon had done a master's degree at the MIT Media Lab, then a PhD while on the team at CERN. He had designed a specific language for specifying the style of a document, called Cascading Style Sheets (CSS), which became a core pillar of web technology in the years to come.

One way to think about style sheets is that the words in this book are the content, while the fonts, the colours and the spacing are the style. Of course, for a web page there are far more varied types of content, and much cooler ways of displaying them. Hakon's solution was to establish a 'cascade' of rules that assigned priority to competing formatting specifications. The writer might input everything in a tasteful, small, grey-on-grey font for a paragraph, but the reader needs to force the browser to make it a large, black-on-white font to actually be readable. So the styling of a document is a battlefield for Ted Nelson's Gods of Literature, the Reader and the Writer.

We needed a testbed for CSS, so we used Dave Raggett's magazine-like Arena browser. Experimenting in this design playground, Hakon, along with fellow W3C programmer Burt Bos, transformed the web into something more elegant and beautiful.

With CSS turned off, a web page is just a series of parts – like paragraphs, headings and navigational links – all down the page with no control over their relative position. Everything is black text on a white background. With CSS turned on, the paragraphs may be moved into columns, the navigation links moved into panels at the top and bottom. Beautiful fonts can be chosen for everything, empty space can be used to good effect, and a colour palette applied. Chaos is given order, and content is given form.

Hakon was far ahead of his time with CSS. It ended up being a cornerstone technology of the web, and when mobile devices later arrived, CSS made the transition smooth. Inevitably, Hakon was a man of many interests, and we couldn't keep him for ever. In 1998, he left W3C to work for the Opera browser start-up; later, inspired by Thor Heyerdahl, he sailed across the Pacific Ocean in a raft. Then he got into orchard-keeping, and in 2017 won the coveted award for Norway's best apple juice.

W3C published the first CSS standard in 1996 and has maintained and updated it ever since. Hakon's work developed new technical ideas and delivered them to the broadest possible range of end users. Of course, not every standard W3C publishes sees such widespread adoption. As Hakon was developing CSS, the consortium put quite a lot of effort into developing another standard for Virtual Reality Markup Language, or VRML. Many of us (myself included) were absolutely convinced that the web was about to go 3-D, and that people would soon navigate the web not by clicking on links, but by moving through 'doors' in virtual rooms.

The 1993 computer game *Myst* had elegantly implemented this concept, although it limited your movement to whatever was on the CD-ROM. Many of us envisioned a conceptual leap to 3-D navigation where you moved from one computer to the next via these virtual doors, just as you did by clicking on a link in hypertext. We put VRML in production in 1994 to encourage people to start building these virtual worlds.

Well, it's thirty years later and virtual reality is still just around the corner. It's possible we were simply too early. The Second Life virtual world was later a sensation, and Mark Zuckerberg has since attempted to implement a similar concept with headsets in his Metaverse. I still think it might happen someday. Recently, we were invited to a design studio in Hammersmith, London, and strapped into a virtual-reality headset. As the headset went live the room I was standing in disappeared and I found myself standing on the surface of Mars. The realism of the simulation was breathtaking, as were some of the 3-D movie clips we saw. We also tried a collaborative work mode, where I could see all the people present in the meeting in the virtual room – as well as a dinosaur that showed up for fun. I hold out hope that someday your computer will permit you to walk through a door and deliver you to an alien world – and, also, closer to home, to speak with remote loved ones as if they are right there.

•

The world that witnessed the birth of the web was a more optimistic place than today. The collapse of the Soviet Union, the end of apartheid in South Africa and the liberalization of the Chinese economy all occurred in just a few years. Democracy was the ascendant political mode, and the globalization of the economy established trade networks across the planet. The 1990s were underway.

The web was the signature technology of this happy time. It generated an extraordinary amount of enthusiasm among first-time users and received almost universally positive press coverage. The roar of a dial-up modem connecting to an internet service provider became the sonic trademark of the early web experience. People would recognize the initial gentle warbling of the 300 bits-a-second modem, but I preferred the harsher, fierce noise when it switched up to 19,200, as it meant content was going to be delivered much faster.

People initially talked of 'cruising' the web, until Jean Armour Polly, a librarian in upstate New York, wrote an article in the local newspaper encouraging young people to visit the public library to 'surf' the web on the new Apple computer the town had purchased. The phrase stuck.

Scholastic Books' legendary guide to surfing the 'net.

Early web culture was so delightful. All sorts of people started to come online. Internet forums, previously hosted on newsgroup or bulletin board service (BBS) nodes, migrated to the web. The experience of hand-coding one's own, personal web page in raw HTML and CSS became a rite of passage for a new generation. The aesthetic of the time was clumsy but whimsical. Content was static, with none of the interactive features a modern user would expect. Data speeds were slow, limiting multimedia opportunities. Video, in particular, was almost impossible. Instead, people used GIFs, which had limited support for simple, frame-by-frame animation.

Early web page authors could be quite creative. When they worried that you would not notice their links, they would add little embedded images of physical buttons sticking out of the page in relief. This 'skeuomorphic' approach helped web readers get over the difference between normal text and hypertext. The button would say 'Click here for more info' rather than just 'More info'. Today, we tend to use a much more minimalist approach with just '. . .' in a slightly lighter shade of grey to indicate 'Click here for more info'.

But there were lots of buttons to press. These were embedded in HTML, along with various 'widgets' which introduced primitive interactive functionality. Many sites had 'counters' tallying how many visitors had clicked through, and some had 'guestbooks' for visitors to sign. Often, web pages would join 'webrings' circularly organized around a particular topic. (Clicking forward or backward on the navigation bar brought you to the next site in the ring.) Since no one ever really finished designing their site, the phrase 'Under Construction' was common, displayed against yellow and black diagonal stripes and flanked by animated GIFs of flashing warning lights.

There was a diffuse, decentralized structure to the early web which I very much enjoyed. As there was no YouTube or Facebook sucking up the majority of the traffic, every site had a shot at the big time. For example, in 1996, Craig Newmark, a computer programmer living in San Francisco, started a web-based mailing list of local events called craiglist.org. He soon expanded the site to include apartment listings, personals, discussion forums and a 'for sale' section. Craigslist was a massive viral hit and expanded to cities all over the world. Even today, it maintains its popularity and its early web aesthetic.

You needed a host server to run your web page, which was challenging for the non-technical person to maintain. Geocities, an early provider of web hosting services, organized 'neighbourhoods' with online 'homesteads' that permitted users to create their own pages with 2 MB of storage. Site designs favoured loud, expressive backgrounds, garishly coloured fonts (often Comic Sans or Papyrus) and eye-straining animations. The site proved massively popular and by 1999, Geocities was the third most visited site on the web, behind AOL and Yahoo. Those were the days.

My invention began to bring me all sorts of unexpected accolades. In 1996, I won a MacArthur Fellowship, sometimes known as a 'genius grant'. That was pushing things a bit, I thought, but I certainly appreciated the money. In 1998, *Time* magazine made a list of the 100 most influential people of the twentieth century. My name was the most recent on the list. The most meaningful response I received, though, was from average users. People loved the web so much. They were just delighted.

What was so good about that time? The fact that anyone could not only make a website, but did – and the fact that most of the web traffic went to all kinds of individual sites: small and medium sites dominated. Researchers call this a 'scale-free network', or a 'fractal' one. With a fractal, as you zoom in or out, you find a similar pattern at any scale. You may recall the computer renderings of the infinitely recursive Mandelbrot that became a craze at one point.

In the late 1990s, the physicist Albert-Laszlo Barabasi proposed that the web (at that time) had a fractal nature. For example, if you were to look at sites related to gardening, you would find a cluster of smaller amateur sites connected by spindles to a few, more professionalized 'hub' sites. Zoom out from the gardening hubs and you'd see the pattern repeated, with the entire gardening cluster connected by spindles to more popular general hobbyist sites. Zoom out again and you'd see the pattern repeated across the entire web. This organic, emergent structure of the early web was a fragile thing of beauty and I was greatly impressed by it. Unfortunately, it doesn't much resemble the web of today.

•

In Geneva, I had been roped into singing in the chorus for the musical *Oklahoma!*, with the Geneva Amateur Operatic Society (GAOS), and had loved it. Soon, I was doing all sorts of things with GAOS. We did a really British Christmas pantomime, for those ex-pat Brits who wanted to show their kids the strange and wonderful aspects of panto without flying them to the UK. I played the dame in *Peter Pan*, a production in which we had six people flying to Neverland, with harnesses and ropes hauled by strong cast-mates. I was the Music Man's sidekick, abetting a con man ripping off the townspeople of Gary, Indiana. We also sang a bunch of things in straight four-part harmony, where a not-very-musical tenor could pick up the notes from the much-more-musical tenors, and hear his voice blend with the other voices to form the whole. I find music relaxing; when I'm stuck on a technical problem I will often gravitate to the piano to improvise.

I was very lucky that my music class at Emanuel School had been taught by the inimitable Mr Strover. He got us singing (I remember the children's cantata *The Daniel Jazz*), and he explained harmony as 'Mr Strover's Gearbox'. His gearbox diagram was always on the classroom wall: it was a set of overlapping three-letter gears, each representing a simple three-note chord. You could start in the 'CEG' slot, playing the C major chord, then connect that 'gear' via the G major chord in the 'GBD' slot above it. (This, in turn, connected to the higher D major gear of 'DF#A'.) Mr Strover got us to hear what it was like to move up a gear, and down a gear, and he pointed out that a lot of songs just stayed in three gears, including much of the pop music repertoire. Thank you, Mr Strover, for your gearbox.

Looking to find a community that might offer a second home for my children, in 1995 I began attending the First Parish Church in Lexington, Massachusetts. First Parish was a Unitarian Universalist church, meaning that not only was my atheism tolerated, but that I was actually encouraged to share this perspective. The chapel was a beautiful old white clapboard structure with a towering steeple – during the American Revolution, its predecessor had sheltered a detachment of Minutemen. My whole family attended and soon I was singing in Benjamin Britten's *Ceremony of Carols*, which the church was staging for Christmas that year. This made the transition from Europe to Boston much easier than it might otherwise have been; what I found at First Parish was a beloved and generous community of people, bound, if not by faith exactly, then by *something*, who were there to help – through thick and thin.

•

About the time I arrived in New England, a former financial analyst named Jeff Bezos was driving cross-country from New York to Seattle with his wife, Mackenzie. Along the way the two discussed potentially lucrative web businesses, settling, ultimately, on a retail bookseller called Amazon. Meanwhile, Marc Andreessen, following graduation from the University of Illinois, departed the cornfields for Silicon Valley. Much of his team from the NCSA followed, and with the support of Silicon Graphics founder Jim Clark, they began to build a follow-up browser to Mosaic called Netscape Navigator. Simultaneously, in late 1994, Bill Gates acquired Spyglass, a small spin-off from the NCSA, with the intention of extending the platform dominance of Windows with a browser of his own.

So capitalism had discovered the web. I had seen this coming, and, in fact, had encouraged CERN's vendors to sell their products via websites from the beginning. What I was not quite prepared for was the frenzy that broke out. Netscape Navigator launched in late 1994 and began to seize market share from its ancestor Mosaic – by mid-1995, it was the leading browser on the market.

Netscape Communications held its initial public offering in August 1995. The sale was massively oversubscribed and the stock skyrocketed in secondary trading. This surprised a lot of people, including me, since Netscape did not have an obvious way to make money. At this stage, the web did not have a way to collect payments: while I very much encouraged people to use the web *for* business, I did not (and *do* not) want the technical infrastructure of the web itself to *be* a dominant monopoly business. But at Netscape, Marc Andreessen had a different philosophy.

Netscape Navigator was free to download, and while in theory Netscape could have charged to use it, in practice I doubted people were going to pay when there were so many free alternative browsers on the market. Instead, Netscape introduced several new browser features, including, most controversially, the cookie, a small block of data about a user's online behaviour that a web server could store on a user's computer. Cookies were a necessary feature – without them, you'd have to enter your password every time you visited a website. But Netscape's cookies ended up like a kind of Trojan horse, invading users' privacy and tracking them all over the web. This would create a lot of problems later on.

Competition in the browser space intensified a couple of weeks later, when Microsoft rolled out its Internet Explorer software. Most people call this the 'first browser war'. (From my perspective it was the second; Viola vs. Mosaic was the first.) Behind the scenes at W3C, Netscape and Microsoft launched into a series of distressing technical squabbles. Carl Cargill, Netscape's representative, was cynical and political – and highly intelligent. At W3C, we had initially adopted the motto first coined by internet pioneer David Clark: 'We reject: kings, presidents and voting. We believe in: rough consensus and running code.' In fact, for many years, the IETF employed something called 'humming consensus' – you could gauge the popularity of a technical proposal by the number of engineers in the room vibrating their lips at the same frequency.

Unfortunately, getting corporate rivals like Microsoft and Netscape to hum the same tune was impossible. The two were pretending to fight about engineering, but, implicitly, they were fighting over their business models. With Cargill's help, we instituted a less vibrational, more process-oriented approach to standards development. I admit this slowed things down – but it also ensured that no one would storm off from the negotiating table.

Cargill was a skilled operator; I still remember a photograph of him and the Microsoft rep with their arms around each other's shoulders. But even as they were pretending to be nice at W3C, the companies were locked in a death struggle in the marketplace. Both began to introduce browser-specific HTML features outside of the consortium. This led to web developers adding the 'best viewed on Netscape' or 'best viewed on Explorer' labels to their sites – again, the threat of fragmentation. The worst-case scenario (for users, at least) was that one of these standards would suffocate the other, leaving the winner with the monopolistic ability to charge fees to access new features on the web.

The Microsoft–Netscape battle also spilled over into the burgeoning field of web applications. These are the devices that make the web interactive, permitting you to fill out forms, watch videos, draw, make music, play interactive games, or anything else you can imagine. Microsoft and Netscape had developed incompatible application scripting languages, called JScript and JavaScript, respectively. They were confusing enough to tell apart on their own but, more confusingly still, neither had anything to do with Java, a third and much more powerful general-purpose programming language developed by Sun Microsystems. Eventually, JavaScript emerged as the common standard; today, it is called ECMAScript and is under the control of the European Computer Manufacturers Association. Meanwhile, web apps have become a dominant force, with many apps written only for the web, and others written first for the web and then ported to mobile systems like iPhone and Android. W3C has rapidly rolled out new features so that the web platform can compete with native apps on the phones.

My preferred solution, of course, was that neither of these sides would win. I began pressing developers to come up with competing browsers, ones less in thrall to shareholders. In 1995, Guido van Rossum introduced Grail, a browser that would let you run applications within itself using his Python language. I met with Guido and encouraged him to develop Grail into a real contender – but, in exchange, he wanted me to start pushing Python inside W3C. I didn't feel I could justify endorsing a programming language which, at that time, fewer than 1 per cent of software developers had ever used. A mistake, as it turns out – today, Python is the most popular programming language in the world.

The good news is that, through W3C, we were able to bring both these sides to the table, preventing a blue text/green text standards war from erupting. Microsoft and Netscape soon realized they were not going to make money off the browser directly – instead, they pivoted to the 'portal' market. This was the regularly updated landing page you were directed to when you first opened the browser – Mosaic's 'What's New' page and Yahoo's 'A Guide to WWW' list were early examples. These portals didn't charge for access, but they certainly got a lot of eyeballs. And that, in turn, lent itself to a huge opportunity for web entrepreneurs: advertising. A new era had begun.

•

When the tech publisher O'Reilly served the first clickable advertisement in 1993, it was essentially just a button. In 1995, though, a start-up called Doubleclick revolutionized web advertising by serving ads through a third-party distribution network. If you had a popular web page, you could allocate some of the space on it to Doubleclick, who would serve a variety of constantly refreshed banner ads to your viewers. The two of you would then split revenue.

For marketers, web advertising offered a radical upgrade over print. The famous saying about advertising was attributed to the marketing executive John Wanamaker: 'Half the money I spend on advertising is wasted; the trouble is I don't know which half.' But Doubleclick could track click-through rates, allowing marketers to target potential customers with greater precision than ever before. The dynamic nature of the web meant advertisers could also make changes to campaigns on the fly. One thing the marketers quickly discovered was that 'ugly' ads worked better than well-designed ones. By 1996, the web was awash in garish advertising, much of it distracting, misleading or just outright hideous.

Many of the ads tried to trick you into clicking on them, by simulating interactivity or 'advising' you that your computer had a virus. The worst were 'pop-up' ads, which hijacked the pop-up function of HTML to clutter your screen with small, annoying windows. Of course, mainstream publishers couldn't ignore the potential revenue from online advertising, and soon there was a rush among legacy news organizations to put their content on the web. In 1996, *The New York Times* began posting the text of its newspaper online for the first time.

The New York Times' *first web page, published in 1996.*

Few understood it at the time, but simple sites like this were planting the seeds for a society-wide shift from paper to digital. From the day the *Times* site appeared, the newspaper printing press began to go obsolete – although there was never quite a 'stop the presses' moment. The transition was as gradual as it was inevitable, and it would take some time before media executives and journalists understood what was happening.

From my time at D.G. Nash, the printer company I had worked at, I had always felt a fondness for print – not just the medium itself, but the actual mechanics of it. At my house in Massachusetts I lived across the road from a classic New England printing press, located in a little industrial area next to the old rail yards. The press ran all sorts of things, although its bread and butter was printing the data sheets for the nearby defence contractor, Raytheon. Of course, when the web exploded, the data sheets moved online, and the bottom fell out of that business. Soon the clackety-clack of the printing press went silent; another machine absorbed by the computer.

Actually, a lot of business models were made obsolete by the web. Because the technology moved slowly and looked a bit amateurish at the beginning, many established firms ignored it. But web technology rapidly improved along new dimensions of functionality and overtook the incumbents. I'm not sure what the first major business line to be disrupted by the web was – perhaps it was the travel agencies. Expedia, founded in 1996 and Priceline, founded in 1997, offered web users the ability to book flights and hotels directly from their computers. This was a faster and far more efficient process, and many travel agents were soon out of work. Expensive stock brokerages were outcompeted by low-cost trading platforms like E-Trade. Many booksellers and mail catalogues fell victim to Amazon, and soon all of bricks-and-mortar retail was under threat. The spread of MP3s, first through websites and later through Napster, foretold the end of many music retailers. While the web created a massive productivity boost in aggregate, the job losses caused by disruptive web businesses could sometimes be disconcerting. I have certainly met a lot of bookstore and music shop workers who resented seeing their customers defect to online stores. (And, with AI, I hear echoes of this debate today, not from booksellers and music shop vendors, but from book authors and musicians!)

Still, it seems that the mega-chains had it the worst. Retailers like Tower Records and Borders are gone, but many local record stores and independent booksellers continue to function. People are nostalgic now, but the megastores were not always so beloved. (In the 1998 web-based romcom *You've Got Mail*, the bad guy isn't Amazon, but a large bricks-and-mortar bookselling chain.) So, in some cases, the web may have given the indies room to compete. I feel it is always good to support small businesses, having grown up near a neighbourhood high street full of real shops.

•

Other stuff was appearing on the web as well, and not all of it was pleasant. Pornography had flooded Usenet since the 1980s, and certainly pre-dated anything I was doing, but the web gave the adult content merchants a new platform. Starting around this time, people would approach me to complain about all the 'horrible stuff' they had found surfing the web. My response to them was not to look at that horrible stuff. There was no technical mechanism on the web that forced people to look at content they found distasteful. If you were looking at something you didn't like, it was because you navigated there: punch something terrible into a search engine, and you would probably get something terrible in return. I encouraged users not to do this, and instead to nurture their bookmarks, and to share good things with others via chat or email.

That was how I felt at the time, at least. Things are different in 2025, as social networks like X employ deceptive algorithms which can *feed* you more and more horrible stuff. (In fact, they thrive commercially off this.) The mechanism doesn't involve direct force

– you can always close X – but it *is* manipulative and coercive. I agree with the historian Yuval Noah Harari, who in his 2024 book, *Nexus*, suggested that when an algorithm specifically boosts something horrible, whoever's operating it should be held responsible.

In the 1990s, the eternal battle between free speech and censorship was taken up by the US government. In early 1996, Congress passed the Telecommunications Act of 1996, a portion of which contained language prohibiting obscenity and indecency on the internet. The courts later struck this provision down, but in the interim it galvanized the nascent online civil liberties movement. A few days after the act passed, John Perry Barlow, the co-founder of the Electronic Frontier Foundation and an occasional lyricist for the Grateful Dead, published 'A Declaration of the Independence of Cyberspace'.

> Governments of the Industrial World, you weary giants of flesh and steel, I come from Cyberspace, the new home of Mind. On behalf of the future, I ask you of the past to leave us alone. You are not welcome among us. You have no sovereignty where we gather.

The Declaration continued in similar language for a thousand words or so and was widely disseminated on the web.

I had mixed feelings about Barlow's manifesto. I liked the idea of the web as an emancipatory political force, but I always assumed – and pointed out – that anything illegal offline was also by default illegal online. In fact, the 1996 Telecommunications Act did something that seemed important to protect blogs: Section 230 contained a 'safe harbour' clause that granted websites and online platforms limited immunity against being held legally responsible for what their users posted or shared. This later proved crucial in the development of social media, as it meant hosts couldn't be held liable for defamatory posts made by users. Of course, now, as social media polarize humanity's opinions, the idea that they should be immune is again subject to question.

Implicit in Barlow's manifesto was the belief that connectivity was all that humanity needed to get along. I knew it was more complicated than that. Politics was not foremost in my mind when I created the web, but I did hope tools I'd built would be used to make connections beyond the standard geopolitical boundaries. By democratizing publishing, and reducing the costs of distribution to almost zero, the web created a flat, open platform in which any voice could be heard.

Now, just because any voice can be heard doesn't mean every voice should be listened to, and the web certainly enabled its fair share of demagogues and cranks. But it also enabled a great number of noble dissidents, who have strived, and continue to strive, to make their voices heard against all manner of oppression. The web has not eliminated authoritarianism, but I believe it remains an unparalleled platform for speaking truth to power.

Also implicit in Barlow's manifesto was the presumption of universal access. That part wasn't true at all – internet infrastructure was not and is not free! You need an internet service provider of sorts (in addition to a computer with which to access it). And, because the internet turbocharges productivity, it can widen the divide between wealthy societies for whom the internet is always 'on' and developing societies who have only limited access to it. This was one of the things I hoped to address with the web, in fact. I thought the internet needed a compact and efficient mechanism to enable delivery to even the most technologically meagre device. (This would later become a key area of focus for W3C.)

The connectivity of the web accomplished something else I really hoped for – it allowed people to make connections across barriers of class, ethnicity and culture. (Language was a harder barrier to cross, although I'm optimistic about developments in instantaneous machine translation.) My vision for the web was always that there might be someone in, say, Accra, with half a solution to a problem in their head, and someone in, say, Vancouver, with the other half. I would call this group collaboration, this *intercreativity*, the web's core principle. Perhaps, with a little compassion, it may yet lead to the resolution of global political problems.

•

By 1996 there were approximately 45 million people using the internet, and for most, the web was the primary point of access. Pre-existing protocols like email and FTP started to be absorbed into the browser, and new users often had trouble telling 'the web' and 'the internet' apart. Modem download speeds were still limited, but an exponentially increasing amount of content was coming online, spread across at least 100,000 websites.

To find things, you initially used online lists. Just as people loved to make a great page on (say) cats, they also loved to make a list of all the pages on cats. Just as it was great to find a new cat picture and put it on your page about cats, it felt great to discover a new web page on cats and add it to your list. That's just how people were. There were lists of lists, of course. At CERN we hosted the WWW Virtual Library, which was a top list of lists in each subject, and we delegated to WWW 'virtual librarians' the task of maintaining lists about physics, biology, cats, whatever. So our virtual library was a delegated hierarchically structured set of lists. But we needed a better way of finding information. Enter the search engine.

Yahoo!, AltaVista, Lycos and Excite were among the many for-profit firms attempting to build a viable search engine in those early days. But search was a hard problem – so hard that it attracted some of the best minds in academic computer science. The preferred approach was to build a bot, sometimes called a 'spider' or 'crawler', that moved from one web page to the next, indexing content and tallying how many times each term or phrase appeared. The problem with this approach was that unscrupulous developers could hack the crawlers by hiding 'invisible' text packed with keywords on the page, simply by matching the text and background colours.

The search engines at the time indexed web pages just as you might index a library of books – simply record the words in each book. They were famously hopeless: people would ridicule the results you would get. And when authors started to manipulate them, they became even more hopeless. Why did the search engines not find the best pages? Because they didn't look at the lists! Or in general, they didn't look at the links connecting sites, only at the text of individual web pages. But, in fact, humans looking for stuff would find lists and links, and they would follow those links to a web page. The problem was that these search engines didn't use the links at all. If you wanted to find a page on jaguars, for example, your best method might be to avoid web-crawling search engines altogether: you would start from lists you trusted to find lists of pages on animals, and then within those lists you would find a link to a page on jaguars. This would work because people carefully made, and curated, lists of links.

Starting in 1997, Larry Page and Sergei Brin, two graduate students at Stanford, began experimenting with a different approach. Called PageRank, their search engine instead looked at the patterns that could be found in lists and the pages they pointed to, and from that structure worked out the pages that people making lists seemed to like. Find the pages that have the words jaguar, sure, but then find a bunch of jaguar-related lists and the web pages those lists point to in order to find consistent clusters of pages, with the same lists pointing to the same pages. In essence, the pages with the best links to them for a given term are likely to be the most salient.*

This turned out to be a much better way than looking at how many times the word 'jaguar' appeared. In fact, the algorithm could spot more than one independent cluster – one about jaguars the animal, one about Jaguar cars, and one about the Jaguars sports team. PageRank uses techniques from linear algebra to determine

* Mathematically, finding all the lists and pages which point to each other in clusters is called a clustering algorithm: you find the eigenvectors of the matrix, and each one corresponds to a different meaning of the word 'jaguar'. Eigenvectors are the way quantum mechanics stipulates which states an electron in an atom can be in. So for a physicist, the results of the search were the quantum states of the whole web of links – an 'oo oo' moment.

the relevance of this particular page with reference to the broader web ecosystem.

I will never forget when I first learned of the PageRank algorithm – I was immediately arrested by its simplicity and its beauty, just as I had been all those years before when I'd heard about the RSA secure key algorithm at Winspit. These two algorithms remain, to my mind, some of the most beautiful ever devised. The venture-capitalist community saw the potential of PageRank at once, and it was soon rebranded as Google.

Another academic problem, and one that we have never completely resolved, is the problem of broken hyperlinks. Projects like Xanadu envisioned a central registry of links, but that required a central overseer, which I did not want the web to have. Still, even early on, I began to recognize we had a problem. What happens when the page you have linked to stops being served? Without a duplicate registry of what was at that destination, the link is lost for ever. In those early days, the web was growing exponentially, and as those interconnections grew, dead links became more and more of a problem. Today, it's estimated that some 60 per cent of all links ever posted to the web lead to pages that aren't there. This affects legal findings, academic citations, and all kinds of other information whose value is diminished by the lack of links.

I am deeply indebted to the archivist Brewster Kahle for providing an important tool in the fight against missing information. In 1996, Brewster, along with Bruce Gilliat, founded Alexa Internet, a for-profit firm that analysed web traffic and offered rankings of the most popular pages. (The name was a reference to the Library of Alexandria.) Brewster realized that many of the web pages he was analysing were short-lived, so, in parallel he founded a non-profit library called the Internet Archive. The archive managed a crawler that took regular 'snapshots' whenever web pages were visited and stored these snapshots in a database.

In 1999, Brewster sold Alexa Internet to Amazon for a reasonable sum. (Amazon later also named its personal digital assistant 'Alexa', but the company claims this is a coincidence.) Brewster turned his attention to managing the Internet Archive full-time, and in 2001 he opened his non-profit database, known as the Wayback Machine, to the public. Brewster has since branched out into archiving all sorts of information, including books, moving images, software and old Usenet posts. A considerable amount of what we know about the early web is due to his tireless work.

Brewster was absolutely a pioneer. Back in the early 1990s, as the web was competing with Gopher, Brewster developed a third internet networking protocol called Wide Area Information System, or WAIS. Like the web, WAIS used a client–server architecture. In fact, WAIS's motto was very similar to my own guidelines for the web: 'Everyone should be a publisher.'

Brewster and I had different design philosophies on how users should access the internet. He was more of a top-down structure guy, who had partnered with large corporations, including Apple and Dow Jones, to make a front end that was more friendly to paid content distributors. His WAIS system was more professional than the web, and had some features the web didn't, including mechanisms for distributing royalties.

Brewster tried to pivot WAIS to make it the default search engine for web content, but by 1995 or so, it was clear that his system, like Gopher, was not going to survive. And yet so great was his passion for digital information that Brewster lost not a second before embracing a new platform. Inspired by a visit to AltaVista's headquarters, he instead became the first web historian. AltaVista, as part of their burgeoning web search efforts, had made an internal duplicate of the web as it existed at that time. The entire web – all the content – could fit on an array of storage devices about the size of a soda machine.

History buffs will recall that the Library of Alexandria was eventually destroyed. Brewster, seeking to avoid a similar outcome for the Internet Archive, decided that this one backup at AltaVista was not enough; he had to start making his own backup of the web. His first effort involved a mechanical device called a 'tape robot', which sorted racks of magnetic digital tape behind glass, and retrieved the tapes with a robotic hand. (CERN had similar tape robots at one point.) The system took up about 2 terabytes of data and held roughly 30 million pages. Eventually, Brewster and his team made numerous physical backup copies of the web, but if you've ever used the Wayback Machine to search for early web content, that robotic hand played a role in bringing it to you. (Sadly, there is no archive of WAIS. Brewster saved my work from destruction, but not his own.)

Today, Brewster and I are good friends. We met for the first time sometime in the early 1990s, and it was obvious we had much in common. Brewster was intense, highly intelligent and a little eccentric – I found him very relatable. 'I think of Tim as the Statesman of the Internet,' Brewster recently said of me.

•

The biggest practical problem we faced was behind the scenes. Following the defection of Marc Andreessen's group, the NCSA in Illinois abandoned maintenance of the popular web server software that had been the counterpart to its Mosaic browser. This software was essential to keeping websites accessible and functioning smoothly. I wasn't maintaining the WWW server software I'd used to first host the website on my NeXT box at CERN either. Instead, a coalition of open-source developers had taken it upon themselves to modernize and update the NCSA's web server kit, through a series of regularly updated 'patches'. Thus the free, open-source Apache HTTP server project was born.

These Apache servers worked really well for hosting individual web pages, but they had a weakness: if a website suddenly became popular, the host would be overwhelmed with requests, causing the page to become unresponsive and the server to crash. I termed this the 'hot spot' problem, although it is better known as getting 'slashdotted'. Slashdot was a website popular in the tech community that allowed users to post and upvote new links – it was one of the first mechanisms for making content go 'viral'. A post that made it to the top of Slashdot might result in hundreds of thousands of users clicking on the link at once, enough traffic to bring down all but the most powerful web servers. In the mid-1990s, I was very concerned that slashdotting would limit the growth of the web.

Now, the advantage of being at a place like MIT is that you will often find yourself in the company of some rather clever people. Sometime in the mid-1990s, Professor Tom Leighton began complaining at a faculty lunch that the web, while obviously useful, didn't really lead to any 'deep' problems in computer science that an academic could study. Offended, I immediately brought up the slashdotting problem, which certainly seemed to me like a worthy challenge for an ambitious academic!

Soon, Tom was presenting the slashdotting problem to a lecture hall full of MIT graduate students. One of those students was Daniel Lewin, a star PhD candidate and wizard in computational complexity theory. In this hall of geeks, Daniel – or Danny, as everyone called him – cut a somewhat unusual figure; he'd spent several years in a special forces unit of the Israeli Defence Forces, rising to the rank of captain. Soon after hearing me present the problem, Danny worked out the 'consistent hashing' algorithm for distributing content requests among a rotating population of web servers. His work on this algorithm won him the departmental award for best master's thesis.

Without consistent hashing, I don't know how back-end infrastructure could have kept up with user demand for front-end content. Hashing solved this issue by reducing the load on any one computer; at the same time, it created a significant business opportunity for whoever managed the server network. In 1998, Danny and Tom incorporated their start-up Akamai Technologies, and the next year they went public. Akamai became one of the leading content distribution networks and is now in the S&P 500 Index. Danny was the original CTO and Tom the original CEO, a position he holds to this day.

Sadly, Danny Lewin didn't get to see his invention flourish. On 11 September 2001, Danny boarded American Airlines Flight 11 from Boston to Los Angeles. He was seated immediately behind Mohamed Atta, the architect of the 9/11 attacks. Danny was killed as the hijackers took control of the plane – probably because he was attempting to stop them. (While confronting Atta, he seems to have been attacked from behind.) This made Danny both the first person killed in the 9/11 attack, and perhaps the first person to die trying to prevent it. At the time of his death, Daniel Lewin was thirty-one years old.

CHAPTER 7

Growing Pains

The systems that would allow the web to grow to any scale were in place. Google provided search, Akamai provided infrastructure, the Internet Archive provided preservation, and W3C provided the standards to ensure the web kept pace with technological change. By December 1998, about 150 million people were regularly using the internet. For most of them, the browser was the primary means of access.

Worryingly, though, that browser was increasingly Microsoft's Internet Explorer, which came bundled with new versions of Windows by default. I was more than a little concerned about that. Bill Gates was late to the web – but once he recognized its potential, he threw the entire weight of his massive company behind an attempt to dominate it. 'The Internet is the most important single development to come along since the IBM PC was introduced in 1981,' he wrote in an internal memo to Microsoft executives in 1995.

Gates's strategy was called 'embrace and extend'. First, he 'embraced' the web – by establishing a monopoly in the browser space. Once that goal was accomplished, he 'extended' the web – by using his monopoly to build new, proprietary standards into the browser, circumventing W3C. As was later revealed in the US Department of Justice's antitrust suit against Microsoft, there was a secret, third component to 'embrace and extend' – what Microsoft execs termed 'innovate', but what the DOJ termed 'extinguish'.

When Internet Explorer debuted in 1995, it was clearly a rush job. But Explorer evolved rapidly, and in 1996 Microsoft introduced ActiveX, a set of software components that allowed users to interact with videos and games directly in their web browsers. There were already plenty of ways of doing this, and the only notable feature about ActiveX was that it wasn't compatible with Netscape Navigator. But Microsoft pushed it hard, leading to the 'best viewed in' phenomenon I described earlier.

Within W3C, it was becoming very clear (to me, at least) that Microsoft was using its dominance in the operating-system market to push Internet Explorer over Netscape. This was worrisome: if Netscape became irrelevant, then Microsoft, not W3C, would control HTML. The tech community hated the idea that anything they created would have to be compatible with Microsoft. Although Microsoft would eventually be called to account for abusing its monopoly, I suspect there were a bunch of projects which failed – or maybe were never even started – because of the anticompetitive situation.

HTML uses tags, like <title>, and this has inspired people to develop similar tags for other applications. But the tag approach, which derives from the old markup language for paper printing, can be a little clunky, so in 1998 W3C published a standard called

Extensible Markup Language, or XML, which was both simpler than HTML and with broader potential applications. Whereas HTML was designed specifically to encode web documents, XML was designed also to handle abstract information and data. XML took off rapidly, with Microsoft making it the foundation of all of its systems, including the highly profitable Microsoft Office suite. An entire industry grew around it, and today XML is used all over the place, including in electronic bank records and fitness-tracking devices.

This was a great step, but I wanted to take it further. I wanted to take the data and allow it to form natural links with other relevant data, encouraging it to encode *facts*. Like, for example, 'Chicago has a population of 4 million.' Once we had enough of these facts, we could even use them to drive logic *agents*. So ask the agent what the largest city in the American Midwest was, and it could use the encoded facts to give you an answer!

To do so, W3C created a language called the Resource Description Format, or RDF. Starting with this language for simple facts, we began to build massive public data encyclopaedias. If you look at locations like WikiData and various public databases, most of it is mere facts – just gazillions of them. RDF gave us a clean and elegant way to represent these facts. Unfortunately, for reasons that remain a bit mysterious to me, Microsoft viewed RDF with deep suspicion. It was very negative about RDF, or anything built on it, and Microsoft managers made it clear to employees that talking about RDF internally would not lead to a promotion. It is strange that the company went down this path. Trying to reverse-engineer their decision-making process, I can only conclude that they must have seen RDF as a threat to Microsoft Office and their other software based on XML.

This divide had a big effect on the web and the industry. XML data looked like a tree – you know, static, hierarchical, with everything in a folder. RDF was more like the web, a dynamic *graph*, where connections might be drawn between any two pieces of data, even unexpected ones. The second approach is so much better. One of the reasons that Google search results are so smart about things like airline flights and facts about people (height, age, title) is that they have built a huge *graph* of knowledge, accumulated from all the data on the web.

Graph versus *tree*: you would not believe how intense the debate over these competing approaches was. The debate was certainly rather technical, but the key thing to understand is that when you encode data in RDF as a *graph*, you make it very easy for a computer to read. And when you do that, you make it very easy for a computer to aggregate. And when you do that, you can combine data from different sources, and get new insights – whether it is fitness and medical data, economic and social data, or data about proteins and genes.

As the web grew, exponentially more data was being added to it every year. What I believed was that by formatting that data into a machine-readable graph, the *semantic* web would be much smarter than the regular web, and it would turbocharge the web's capabilities. The day-to-day mechanisms of trade, bureaucracy and our daily lives could be handled by machines, talking to other machines. And by enabling these machine agents to understand the web, we would, in time, help to bring about a revolution in artificial intelligence. To me, this is what this debate was really about! With RDF, we were trying to get our computers to *think*.

The idea that large, structured data sets could fuel the development of AI will now seem obvious, but believe me, in the late 1990s almost no one was thinking along these lines. My closest ally in pushing for the semantic web was Jim Hendler, a professor at the University of Maryland who was attempting to wrangle the very large amounts of intelligence data that the US espionage apparatus gathered each year. Hearing of his interest in semantic systems, someone in the Pentagon invited Jim to Washington, DC. They met at the Defense Advanced Research Projects Agency facility, where Jim was shown a sample of the huge number of photographs of Russian-built MiG fighter aircraft the US government had collected. There was data coming in from all over: from aerial spy planes, from satellites, from sources in the field. Looking for a way to manage it all, they made Jim a programme manager at DARPA for the next couple of years, from where he promoted semantic web development across labs in the USA.

Jim was friendly, easy-going, big, cuddly, and basically, well, huggable. Defence contracting was not my thing, but I saw the opportunity to use this work in a much broader way. With Jim's help, we were able to convince DARPA that whatever specification we came up with would be public, open-source and royalty-free. Jim and I were later joined by Ora Lassila, a semantic web enthusiast who worked for Nokia in Boston.

Unfortunately, as Microsoft continued to 'embrace' the web, its influence began to rival that of W3C. Internet Explorer won the first browser war in a rout – by 2000, it was approaching 90 per cent market share. (Netscape, meanwhile, sold itself to AOL, which subsequently merged with Time Warner, in what is now regarded as one of the most disastrous corporate mergers of all time.) Of course, the next phases were to 'extend' the web by forcing de facto Microsoft-friendly standards on users, then 'innovate' (or, in the DOJ's terms, 'extinguish') by deprecating competing technologies. That's what happened to RDF, which, in the early 2000s, entered a period of remission people call the 'semantic winter'. In later years, though, the semantic web would become important once again.

•

As Microsoft's ambitions climbed, I did too, though I was scaling a mountain of a different sort. The Black Mountains across the south-east of Wales have continued to call me back since childhood, particularly when I've felt the need to contemplate tricky problems.

Whether dreaming up new things or just relaxing over a break, green space and green places have always been very important to me. When we were small children, we would go on family trips to the wildest westerly parts of Cornwall, and then, when we were older, we would go to rural Wales. We would travel to the same spot, stay in the same little stone cottage, every year. And then, when our children were growing up, we would take them too.

The cottage is located in the Black Mountains, the first serious hills across the border from England, and once you are there, no other houses are in sight. The stream that flows down the hill past the door is diverted into a drystone-built culvert and through a stone tongue. It collects in a pool for washing – a sort of shrine to water. That basin is the focal point – that's when you know you have arrived. But you aren't really there until you have gone up the hill itself.

Pick a path – perhaps the wide grass one, curving gently through the forest before breaking out into the sheep fields and sheep pens, then onwards to the moor of heather and wind-blown hilltop. There, you can see and understand the system of valleys you are in. For the physical investment of just a modest climb, you can now walk ridges for hours. Do you hike immediately all the way to the northern escarpment, or leave it for another day? Maybe you chat. Maybe you swim in a river, maybe you picnic by a lake. And then you return home, all together.

In his memoir *Waterlog*, Roger Deakin writes of swims he took all over Britain, but also of the magnificent bath he ran at the end of one particularly big day. That bath is magnificent because of its antique workings and its depth and its steam, but the bath in our cottage is magnificent for another reason: simply because I put in every inch of the shiny copper pipe necessary for its log-fire heating with my own bare hands. I can pull the plug and hear the water gurgle away – with no leaks. The familiar, happy sounds of family and friends percolate through the thin walls. There is supper in front of the wood stove. There will be discussions of everything over dinner. There will be plans for the next day. There will be discussion of everything else over the washing-up.

It is a grounding place, both for retreating from the hard, pressing problems of life and to talk them over and to navigate a way through them. When I am in a safe harbour, my mind subconsciously sifts through all that is going on around me.

•

Of course, as I considered the semantic web and this arcane battle for behind-the-scenes control of logic languages, almost no one else was paying any attention. Instead, they were watching the dot-com bubble. When Netscape had conducted its IPO in 1995, I thought the company was overvalued, but after Marc Andreessen and Jim Clark flipped their company to AOL for $10 billion, an enormous number of copycat businesses – light on profits, heavy on marketing – entered the fray. At the same time, discount online brokers like E-Trade and Ameritrade encouraged retail traders to speculate on these ephemeral companies, driving up their thinly traded stocks to absurd valuations.

The standard-bearer for the dot-com bubble was Pets.com, an online retailer of pet food that spent a fortune on a Super Bowl ad featuring a sock puppet, while struggling with other, more basic business tasks, like fulfilling customer orders. Pets.com went bankrupt in 2000, but there were hundreds of similar businesses, each more absurd than the last. Somewhere around this time my financial adviser informed me of an online cattle market that was briefly worth more than the entire cattle industry. I understood at that point that dot-com stocks resembled the tulip mania of the seventeenth century, and strongly discouraged my friends and business partners from investing in these businesses.

The Pets.com phenomenon also led to frenzied speculation in the names of dot-com addresses. This was mostly due to my innocent decision to harness web technology to the domain name server infrastructure way back in 1990. I could not have predicted it at the time, but in the late 1990s the leadership of DNS fell into the hands of some rather unscrupulous operators. What had been an unremarkable piece of back-end internet infrastructure was temporarily transformed into a revolting for-profit bonanza.

As the web grew in the mid-1990s, speculators called 'cybersquatters' began to buy up DNS addresses like 'restaurants.com' or 'printers.com' in the belief that these domains might one day be valuable to retailers. However, some of the more enterprising cybersquatters started targeting names of actual, pre-existing companies. Sadly, DNS's leadership sought to profit from the trend. The DNS had, for many years, been administered by Jon Postel, a computer scientist who had worked alongside Cerf and Kahn in the earliest days of networking protocols. Postel had long hair, a long grey beard and thick glasses – a real wizard. As tended to happen in those days, his ad hoc custodianship of the early name registry for ARPANET (the initial DARPA-funded internet) evolved into a kind of benevolent dictatorship as the technology expanded. By the 1980s people were calling Postel 'the God of the Internet'. If two people had a dispute over who owned the domain 'thread.org', for example, Postel was trusted to sort it out. With the web's exploding popularity, though, the workload was becoming too much for Postel to handle. (Postel, tragically, died of complications from heart surgery in 1998.)

His legacy was betrayed in the early 1990s when, instead of creating a non-profit organization to run the internet in the public interest, the US government auctioned off the dot-com side of the DNS registry to a for-profit operator. What had been one man's purview was transformed into a cut-throat market, and Verisign, the operator, was later able to make a huge amount of money charging rent on domain names. It didn't have to be this way; the '.edu' domain was run by a public interest organization and worked perfectly well. So did the '.gov' and '.mil' domains, where the US government retained control.

Not all the dot-coms were flops – Google, Amazon and Netflix got their starts during this period. And while most of the flimsy start-ups faded, the excitement of the era resulted in many people purchasing goods online for the first time. Just as web advertising and publishing transformed the media business, web-based e-commerce began to transform the retail shopping experience. (As with the media, it took some of the legacy bricks-and-mortar businesses a long time to understand what was going on.)

The popularity of the web was accelerated by massive improvements in the home computing experience. Dial-up modems were replaced by broadband internet connections, improving data throughput by an order of magnitude. Clunky CRT monitors were upgraded to elegant liquid-crystal flatscreens, rendering images and video with greater fidelity and clarity. Processors, following Moore's Law, continued to pack exponentially more transistors onto the silicon chips, dramatically expanding the PC's graphics capabilities. Behind the scenes, teams of developers competed to establish new methods for compressing multimedia data, resulting in an alphabet soup of new formats: JPEG, GIF, AVI and MP3. Due to this immense and (mostly) hidden engineering effort, the personal computer evolved from the disconnected, limited-use workstation of the 1990s into the networked multimedia platform of the 2000s.

Keeping ahead of these developments at W3C required an immense amount of effort.

We moved from HTML to HTML2 in 1995, to HTML3 in 1997, and to HTML4 in 2000. Each new HTML standard tried to 'skate where the puck is going', by anticipating the rapid development cycle of the personal computer. At the same time, W3C looked, when possible, to supplement existing proprietary standards with free, collaborative formats of our own, like with Portable Network Graphics, or PNG, an image-encoding standard that W3C developed as a free, open-source alternative to proprietary image-encoding format GIF.

PNG development at W3C was led by Thomas Boutell, but the finished recommendation listed the names of more than twenty authors. Most of these authors were volunteers. Multiple working draft revisions were required to make the standard work, following the consensus working-group model. PNG worked well and was, like all W3C standards, completely royalty-free. This combination proved popular, and it rapidly became one of the most widely used image formats on the web.

The arrival of the multimedia PC coincided with an upgrade of the development environment at W3C. Microsoft Exchange Server had pushed me to the brink of madness several times, so it was with great relief that we switched over to Mac OS X shortly after its arrival in 2001. In 1996, Apple had purchased the remains of NeXT, and soon Steve Jobs was reinstated as the CEO. As a long-time NeXT enthusiast, I could easily see the platform's DNA in the new Macintosh operating system. I have used it ever since.

Following AOL Time Warner's massive $50 billion write-down in 2001 – the largest in the history of business up to that point – Netscape began to dwindle away. The web bubble popped, a great number of start-ups failed, and the retail dot-com speculators returned to their day jobs. But it was in this somewhat deflated environment that perhaps the fullest realization to date of the original idea for the web was born: Wikipedia.

My original WWW client didn't take you to a portal; it took you to what I called a Home Page. The home page was where you started your exploring, where you kept links to interesting things, and to lists – editable lists of all your favourite places. It resembled bookmarks in today's web, but in editable hypertext.

You could make a blank page with a blinking cursor, type into it, and link it to other parts of the web relevant to your life. The idea was always to have users create as easily as they consumed. Viola, Cello, Netscape, and the browsers which followed them, alas, did not allow editing, and so the word 'home page' ended up being used for an organization or person's main page.

Wikipedia, which launched in January 2001, had a similar philosophy, inviting everyone to jump in and start editing the encyclopaedia. The rapidly editable Wiki pages were invented in 1995 by Ward Cunningham, an American software consultant. The technology was niche; many regarded Wikis as a toy, including Cunningham himself. He didn't patent the concept because it 'just sounded like something that no one would want to pay money for'.

The Wikipedia project grew out of Nupedia, a failed attempt to create an online encyclopaedia using peer-reviewed expert sources. The experts moved slowly; in the first two years of the project, they published only a couple of dozen articles. Frustrated by the pace of development of Nupedia, Jimmy Wales and Larry Sanger, the site's co-founders, encouraged the experts to use Wiki technology instead. The experts, appalled to have amateurs reviewing their work, rejected the proposal.

But the public loved Wikipedia. The site grew rapidly, adding hundreds of articles a day. Soon the project was adding images and video, and moving to other languages besides English. Articles started out primitive, and often contained inaccuracies, but rapidly evolved as collaborators from around the world corrected and proofread them. Wikipedia has grown to contain millions of articles on every subject known to our species – an invaluable repository of human knowledge that I consider one of the modern wonders of the world. What made this system work was *intercreativity* – a group of people being creative. Wikipedia is probably the best single example of what I wanted the web to be.

CHAPTER 8

Technology and Society

As the web grew, politicians began to take an interest in the technology. Some wanted to pass laws safeguarding users' privacy; often, though, they wanted to force web hosts and internet service providers to share users' data with law enforcement and state intelligence agencies. While I respected the need to go after cybercriminals, I felt it was important that web users had at least a default expectation of privacy.

A key actor in this drama was the Electronic Frontier Foundation, a non-profit organization that was created in 1990 to fight for rights on the net by Mitch Kapor, John Gilmore and John Perry Barlow, the author of 'A Declaration of the Independence of Cyberspace'. The EFF's deputy director, Danny Weitzner, soon left to co-found the Center for Democracy and Technology (CDT), which in turn joined W3C to add its weight to the drive for civil liberties on the web. At one point we were looking for a leader for W3C's Technology and Society domain, and Danny, on behalf of CDT, was very keen that the right person should get the job. I soon realized that his own profile was ideal. In 1998 he started a long and productive stint in the job. One crucial contribution was to steer W3C through a potentially disastrous incident over patents.

In the early days at CERN, we had managed to get a commitment that the organization would never charge royalties for WWW technology. We had thought that this had firmly established a free web. However, we were wrong. Just under a decade after the first website, a member of a W3C working group announced that the standard he was working on would incur a royalty fee. We were shocked: the implied threat was huge. Once the culture of openness was broken, all bets were off. The big patent holders out there, including IBM and Microsoft, would descend on every aspect of the web. That would have been the death of the free web as we knew it – and possibly of the web itself.

W3C had to prevent groups being blindsided by announcements of previously hidden patents. So a group was formed, led by Danny, which produced W3C's patent policy. It was adopted in 2003. The big companies realized that, by backing off from patent enforcement, they might not get a royalty share of the cake that was the web, but that the cake itself would grow much bigger. The W3C patent policy made W3C a fairer place to create standards. It is no exaggeration to say that it saved the web.

•

Along with the Electronic Frontier Foundation, W3C would also protest – and code – for an open web with no government spying. Of course, it was one thing to say that web traffic should be free from limitless government snooping. It was another thing entirely to build the protocols to make that possible. Fortunately, at MIT, I was lucky enough to work in the same building as Ron Rivest, the brilliant cryptographer who was among the creators of the beautiful RSA secure key protocol that had so impressed me many years earlier in Dorset. It would be just the ticket for creating more secure identities on a more secure web. The MIT Computer Science and Artificial Intelligence Lab (CSAIL) held a faculty lunch every Thursday in a large lab conference room. During one of those lunches, I gave an impassioned speech in favour of the RSA system, in response to the IT people's threat to move the institute from RSA keys (secure) back to passwords (insecure). 'This is a gift from mathematics,' I said. Ron came up to me afterwards and thanked me. (He later told a CSAIL historian he'd come up with this elegant concept after drinking too much Manischewitz wine while celebrating Passover.)

Ron, Danny and I became good friends. The lab organized a group trip to the Wequassett Resort in Cape Cod every year. Danny and I would often take to puttering – or zooming – around Wequassett Bay in a small Hobie catamaran. This became a tradition, regardless of whether there was wind or not.

An important step for security was agreement on the HTTPS standard in 2000. (The 'S' stood for Secure.) HTTPS encrypted all outgoing and incoming hypertext traffic using the RSA algorithm. I was very happy with the way the two systems intermeshed; although they'd been designed in parallel, the World Wide Web and public-key encryption turned out to be a perfect fit. Within W3C, we began pushing everyone to upgrade from regular HTTP to HTTPS, thereby concealing sensitive information like bank data, medical records, passwords and private correspondence from prying eyes. We hit an unfortunate snag when we changed the URL to add an 'S': it broke all existing links on the web! In retrospect, the web would be much more functional and simpler if HTTP and HTTPS pages both used the same URL, just with different protocols.

•

I really had Michael Dertouzos to thank for making all this possible. He organized the lunches and the Wequassett visits, and he was the one who'd brought all these incredibly talented people together at MIT in the first place. But Michael felt, as I did, that technology should always serve the human being, not the other way around. 'We made a big mistake 300 years ago when we separated technology and humanism,' he once told *Scientific American*. 'It's time to put the two back together.'

So I was heartbroken on 27 August 2001, when Michael died, following complications from surgery. He had been my friend, mentor and main supporter, as he had been for a great many others at MIT. Ten of us flew to Athens in Greece for his funeral. On the flight there, I was seated next to Jean-François Abramatic, who'd been elected W3C chairman in 1996. We flew right over the Parthenon, and the two of us peered down out of the window to the Agora steps where Socrates once conducted his dialogues. I remember our own dialogue quite well.

'This is where our civilization started, but they couldn't have imagined the internet and all the stuff we've got now,' Jean-François said. 'Isn't it great we got from their beginnings to an internet-based society?'

'Well, but there's been a few hiccups along the way from Athens to here – the sack of Rome, the Visigoths, the dark ages,' I said. A question occurred to me. 'Do you think in our future there will be another dark age, or have we got to a point where civilization and liberal democracy are locked in?'

'I think it's locked in,' Jean-François said. 'Now liberal democracy is so strong, and growing in concert with technology. I don't think we'll have another dark age.'

The plane landed and we attended the funeral. It was a sombre affair, but gave us the opportunity to offer our recollections of Michael and the many great ways in which he had helped so many of us. The funeral occurred on 5 September 2001. Afterwards, we returned to Boston. A few days later, we found ourselves in the American Academy of Arts and Sciences, headquartered in a Frank Lloyd Wright-style pavilion in Norton's Woods, Cambridge, Massachusetts. We were having a W3C strategy session in their beautiful library when we heard about the 9/11 attacks. Turning on a news stream, we watched in horror as the towers went down. Jean-François was standing beside me. He looked at me and said, 'You know what? I take it back – about civilization being locked in.'

•

The continuity of civilization had been on my mind a fair amount that year. In 2001, I was elected a fellow of the Royal Society. It was a very great honour for me – many of my scientific heroes were Society fellows as well, including Alan Turing. I was presented to the Society at an induction ceremony, then brought into the office to sign the registry. At the top of this 350-year-old leather book I saw the signature of King Charles II, followed by the signature of Isaac Newton. Paging through the registry, I passed thousands of signatures, including those of Benjamin Franklin, Charles Darwin, Albert Einstein and Stephen Hawking. At the bottom, with a sense of humility, I signed my own name.

The Royal Society is a pretty useful institution. Its headquarters are located in Westminster, London, in a collection of white stucco houses facing St James's Park. As a fellow, you can just go in and work there whenever you want. The view is splendid, and the wifi is decent. It was as good a place as any to reflect on my discussion with Jean-François.

How intertwined were democracy and technology, anyway? Did they necessarily grow in concert? My assumption had always been yes, but my work with Danny Weitzner had shown me that it was not enough simply to release new technology and hope for the world to improve. You had to develop technology and society together. You really had to fight, in a principled and continuous way, for human rights. The web offered people a platform for their voices to be heard, reducing the cost of publishing and distributing information to effectively nothing. But, used improperly, it could also be turned into a tool of surveillance and control.

The printing press, when it arrived, had allowed the distribution of a great number of new ideas. Those ideas challenged authority and established norms, leading to centuries of political and social unrest. In Western societies, that foment of ideas eventually stabilized into the prosperous liberal democracies of the late twentieth century – but that experience was not universal, and perhaps it was also not inevitable. In fact, in the wrong hands, new communications technologies could do as much harm as they did good. This was especially true of mass broadcast media like radio and television, which often became tools for authoritarians. The web offered a new frontier – it was like giving everyone their own printing press, everyone their own radio transmitter, everyone their own television station, all with no oversight, all at once. To me, that *had* to be a good thing. I believed, and still believe, in the emancipatory power of the internet.

But I also had read enough history to understand the potentially destabilizing effect this shift might have. Democracy hadn't arrived in a neat little package; it was the product of bloody revolutions. The term 'human rights' was itself a product of the independent press; it could be traced to self-published US abolitionist literature in the early 1800s, among other sources. And, of course, a slogan wasn't enough to get rid of slavery. A war had to be fought.

By the early 2000s, I was beginning to see that the web was not merely a rival to print, television and radio – it was going to decimate and supersede those media. This was perhaps clearer to me than it was to executives within the media industry. I remember being interviewed by the BBC's Alan Yentob around this time. As I like to do, I decided to turn the tables and interview *him*. Was the BBC aware that in a few years, monolithic broadcasting would give way to streaming channels? Did the BBC have a strategy in place for that? Yentob, surprised, said he would investigate the issue.

It was also obvious, even in the early 2000s, that the web was going to challenge existing authorities and established norms. I did not want to usher in a new age of unrest, but I did want to make sure the web had the ability to reach and uplift everyone on the planet. If everyone had access to the web – really, *everyone*, not just the technophiles or the citizens of wealthy countries but also the poor, the repressed, the voiceless and the invisible – I felt it would help democracy to flourish.

•

Somewhere in the early 2000s, the one billionth web user came online. It was among the most rapid adoptions of any technology ever, but it was still a challenge for many people to access the web – whether due to location, poverty or disability. Within W3C, I pushed to make accessibility a priority. Fortunately, I had the assistance of several extraordinary individuals.

Judy Brewer joined the W3C team in 1997 to lead development of our accessibility standards. Judy is an electric-wheelchair user and a tireless champion for people with disabilities. She has a rare metabolic disorder, which she addressed in head-on fashion by securing a PhD in molecular biology. Travelling with Judy, I witnessed personally how poorly airports are set up for people travelling in a wheelchair, especially getting electric-wheelchair batteries through security. It's a disaster.

Early on, Judy and I decided that any spec W3C produced would be reviewed for accessibility. This surprised certain members in the consortium, but it also motivated them, and I came to feel this decision was one of the best I made as the W3C director. The accessibility community brought a tremendous amount of enthusiasm to the web, because of the web's ability to channel the same content through multiple different formats. With the right standards, we could ensure you could click to transform any web page into a large-print format, switch schemes for colour-blind people, display alternative text for images, or even have the page read aloud. The web was streets ahead of other media in this regard.

Our accessibility work was also greatly helped by Shadi Abou-Zahra, who joined W3C in 2003. Shadi was an energetic and garrulous software engineer who also used a wheelchair. Studying computer science in Austria in the 1990s, he'd experienced great difficulty navigating the lecture rooms of his university. He'd subsequently become a leading representative for disabilities advocates online. When we first met in Boston, I liked him immediately.

Shadi has observed that a good accessibility standard can benefit a very great number of users. For example, as we travelled round Boston, Shadi would often point out the availability – or, more frequently, the lack of availability – of kerb cuts or dropped kerbs on pavement corners. These small ramps allow wheelchair users to move about the city freely, and are also used by all kinds of people: families with pushchairs, kids on scooters, even workers pushing heavy loads on wheeled dollies. What opponents considered a costly concession to a special-interest group was in fact an affordable benefit for the whole society.

Another example the W3C accessibility folks often used was closed captioning on television. This, too, had started as a push for accessibility, with work starting at the Gallaudet University for deaf people in the 1970s. The TV networks had originally resisted it, but as it turned out, a wide range of people loved closed captioning. Closed captioning allowed you to watch television in loud environments like bars or airports; it allowed you to interpret accents you weren't accustomed to; it allowed you to watch TV with the sound off if your partner was sleeping next to you; it allowed you to follow along with the plot if the script was complicated. Later, closed captioning would even allow web users to screenshot funny dialogue from television and movies and transform them into shareable memes – a use case the pioneers at Gallaudet never could have envisioned when they first advocated for the technology.

Shadi has noted that the words 'accessibility' and 'disabled' sometimes carry a stigma with them. 'My mom, she's in her nineties, but even today, she would never say she's disabled,' he told me. 'She only hears less, and sees less, right?' For this reason, Shadi sometimes also calls the guidelines 'ease-of-use' standards. The work by the W3C accessibility team has really made the web a far better and more inclusive space. Features like alternative text for images, or pronunciation guides for new words, don't help just those with disabilities but can enhance the presentation of the original content for everybody. Contrast improvements, text resizing and alternative colour presentation schemes – all originally intended to help those with visual impairments – are also great for preventing eye strain.

Like closed captioning, some of the Web Accessibility Initiative's work now intersects with AI technology in ways we did not expect. One of the best examples is machine translation. We've grown accustomed to asking AI to automatically translate between languages for us, but one thing W3C has long pushed for is alternative simplified text *in the same language* that replaces jargon with explanations at an elementary reading level. While I sometimes encounter knee-jerk reactions to 'dumbing down' the text in this way, simplified text can open difficult material to a much wider audience: non-native speakers, young people and even the newly literate. AI-simplified versions of text can help people navigate complicated medical directions, correctly interpret government forms, and understand arcane language in business contracts. By automatically generating simple, plain-language versions of things such as credit card offers, immigration papers, military enlistment contracts and legal documents like wills, we generate great social benefits and ensure fair dealing. Simplified text even helps preserve cultural heritage – archaic writers like Shakespeare and Chaucer can be made much more accessible by layering simplified, modernized versions of their work side-by-side with the original.

A few summers ago I ploughed through Neal Stephenson's novel *Termination Shock* as a book and as an audiobook, switching between the visual and audio forms of the text where necessary – word by word at times. Reading it in company, listening to it when alone. Reading it when a passenger, listening to it when driving. I found this to be an effective way to experience the book, and I'd like to apply the approach to all kinds of content, including blogs, podcasts and newspapers. It is coming. There are apps that can assist, but the operating systems on devices should up their game to provide this kind of functionality automatically. Fortunately, AI can make automatic conversion between text and audio easier.

Automatic text-to-speech services have long been an important goal for disability advocates looking to aid people who are dyslexic or visually impaired. Now, with AI, we can automatically generate speech from text, which is useful not just for blind people, but also for other commuters or gym-goers looking to listen to news articles or audiobooks generated on the fly. Similarly, the accessibility group at W3C has long pushed for audio descriptions that 'narrate' the visual action in movies and TV. AI should allow us to be able to generate these descriptions automatically – and not just for scripted content, but even sporting events and live video feeds. The benefits, I imagine, will extend to everyone.

Shadi's work has ensured the web can reach a much wider audience than ever before, connecting it not only with the disability community but also with young people, older people, non-native English speakers and even technophobes. 'When we do usage studies, we typically find that while around 10 per cent of the population *needs* these services to interact with the web, up to two-thirds of users *benefit* just from having them available,' Shadi says.

•

In the 2000s, as the web expanded, W3C began to open satellite offices in regions all over the world, including locations in Australia, Finland, Morocco, India, Brazil and Senegal. In 2006, we opened the first satellite office in China – at Beihang University in Beijing. Angel Li, our business manager, started there fresh out of study. Angel wasn't a technical person – she'd studied art history – but she shared my vision for a free and open web and pushed hard to make it a reality.

When I invented the web, in 1989, China had no internet access whatsoever. (China got its first IP address in 1994.) But by the mid-2000s, following an extraordinarily rapid period of industrialization and urbanization, it was clear that China would soon be the world's single largest web domain. Angel saw this as well, and she pushed for W3C to establish free and open standards in China for everyone to use. The value of integrating the Chinese internet with the global community was obvious, and with a population of over 1.4 billion people, the Chinese market might double global web traffic all by itself.

China also provided a completely new perspective and set of values – one that could be challenging. Angel served both as my interpreter and guide in navigating this new territory, which was utterly foreign to me. With her assistance, Rosemary and I visited Beijing in 2008, to lecture at the Great Hall of the People, the immense government building which hosts the annual congress of the Chinese Communist Party. The space was enormous, draped in gold and red, with giant concrete walls curving around to either side. I stepped out to face an audience of 10,000 people, one of the largest crowds I've ever addressed. Remembering Michael's advice, I'd written ten slide titles on the back of an envelope and presented, through a translator, a lecture on the future of web applications.

On a subsequent trip, I also met Jack Ma, the co-founder of Alibaba. Ma had a high profile at the time, and it was a big meeting, with lots of photographers present. I liked him – he was very personable, smart, outgoing and thoughtful, and he was not afraid to speak his mind. Alibaba was a kind of industrial version of eBay, connecting Chinese factory cities like Shanghai, Guangzhou and Shenzhen with customers all over the world. If you needed 10 tons of rolled steel shipped to Kuala Lumpur, or a container full of blank T-shirts delivered to San Diego, you bought it through Alibaba.

The rapid industrialization of China had no precedent in human history. It was due, in part, to government investment, and, in part, to the government's embrace of a market economy. It was especially due to the hard work of millions of factory labourers, often working long hours for low pay in dangerous conditions. But the web played a role too, efficiently and rapidly matching Chinese manufacturing concerns with customers all over the world. Acting as a kind of interface for globalization, the World Wide Web facilitated a tremendous amount of economic growth within China – it helped to lift an enormous number of people out of poverty. This was something I could never have imagined happening while tinkering with code in Switzerland all those years before.

China adopted the web with huge enthusiasm. Even as Alibaba connected China to the rest of the world, new businesses arose to serve the country's vast and newly prosperous internal market. The tech companies Baidu, Tencent and Xiaomi soon rivalled US counterparts for size and influence. The W3C Beijing office grew dramatically, until, with Angel's help, it became the fourth international W3C host in 2013, joining Japan, France and the US.

In 2010, Google exited the Chinese market, after failing to reach a consensus with the government on whether to censor or filter out sensitive topics from the search results. The more compliant Baidu took over, and over the following years, the initial promise of the Chinese web has faded somewhat. The government has become more censorious, and the original generation of web entrepreneurs were brought to heel before the state. (Jack Ma had shown the temerity to criticize the Chinese banking system in 2020; he went missing for a while after that.)

Of particular concern is the Chinese government censorship apparatus termed the 'Great Firewall of China'. The purpose of this vast technological filter is to limit access to foreign websites that might present alternative narratives to the Chinese Communist Party's official version of events. For example, searching for the phrase 'Tiananmen Square 1989' in the Chinese search engine Baidu will return no mention of the protests and massacre that occurred. Many other politically sensitive topics are censored as well: the struggle for independence in Tibet, the political status of Taiwan, information about the Falun Gong religious movement, mention of repression of Chinese Uighurs in Xinjiang, even basic biographical information about Chinese leaders like Xi Jinping and Deng Xiaopeng.

Today, the web in China is far from the open and independent platform I would hope for. Still, I'd say it's an improvement on the alternative of nothing at all – the web remains a critical engine for economic progress, and independent voices can still be found there. I remain optimistic about the potential of the Chinese web.

•

In Massachusetts, several W3C employees lived at one point in an enormous five-storey Victorian house in Somerville we called 'The Ranch'. The place was rented from a tech entrepreneur who had installed a big, fat, T1 internet pipe in the cellar alongside a gigantic router. The house was quite beautiful and had all sorts of features: a wood-panelled sauna, a bar and an incredible basement pool with a domed mosaic ceiling, decorated with geometric Arabian stars.

Over the years a lot of people lived and hung out at The Ranch, and they also hosted guests from W3C's satellite offices there whenever they came into town. In this way, it came to reflect the culture of W3C, along with its employees: an incredibly diverse group of talented, energetic and imaginative individuals. Some of the residents made improvements to The Ranch. One of our engineers, Sandro Hawke, had his private telephone exchange installed there. Another, Eric Prud'hommeaux, built a zip line off the roof that connected to a nearby tree. Finding your way to Eric's room was easy in a sense because someone as a prank had painted a wide asphalt-grey path all the way from the front door to his bedroom, with the standard US double yellow line up the middle. Sadly, the original owner eventually moved back in, and they no longer had access to The Ranch. I don't know if the line was ever painted over.

The W3C culture of hardware and software geekery wasn't limited to The Ranch; we loved to come up with new technical solutions to everyday problems. In the days before internet conferencing, you had to have a physical telephone 'bridge' for everyone to call into. W3C had one that connected into the MIT digital telephone exchange that we named Zakim after Boston's new highway bridge. Ralph Swick, the long-time COO of W3C, connected a small computer to watch and control the bridge. On it, Ralph programmed a chatbot so that we could all chat to 'Zakim' directly. Zakim ran most of our meetings, could identify and mute noisy people, and it could even take minutes, queue agenda items and tabulate the results of voting and polls. Finally, when a virtual meeting concluded, Zakim would, on request, publish the minutes online. (Our philosophy at W3C was that if it wasn't on the web, it didn't exist.) In many ways, Ralph epitomizes the culture of W3C, and he effortlessly connects the worlds of software and hardware. He lives quite far out in rural Massachusetts, and rescues people in his big orange emergency vehicle whenever a serious storm rolls in.

One of my professional highlights each year was W3C's annual Technical Plenary Advisory Committee conference or TPAC, as everyone called it. It brought together around 1,000 of the brightest minds in web development in a different location every year. These were people struggling with the most difficult issues on the web, but they were also there for each other. The atmosphere when they were together was always electric, constructive and fun. These were my people. These were the web's people.

We'd travel to Madrid or Shenzhen or Santa Clara and participate in a week of working-group meetings, panels and breakout sessions to resolve the most challenging technical or social issues facing the web. Amy van der Hiel, my personal assistant, with whom I worked closely for over twenty years, had a background in art museum curation and so was well suited to being in charge of the planning. She understood the technical arguments, the people and the philosophy behind W3C. She was able to be both a sounding board and someone who brought people together to help us reach a conclusion, by quietly understanding the essence of technical differences and their particular individual views and opinions.

Not that everything was always so smooth inside the conference rooms. Close to 400 dues-paying organizations were represented at the usual TPAC conference, and they had a widely disparate range of views and objectives. As you might expect, the representative from Google often had a different point of view from the representative from *The New York Times*. The representative from the Wikimedia

Foundation did not always agree with the representative from Microsoft. The points of common agreement between AT&T, Cloudflare, the University of Illinois, the Chinese Academy of Sciences, Fraunhofer Gesellschaft, the BBC, Hitachi, El Instituto Tecnológico de Costa Rica, Taiwan's Ministry of Digital Affairs, the National Library of Sweden, ByteDance, Salesforce and the US Department of Homeland Security were sometimes limited. But all were dues-paying members of the consortium and were entitled to a voice. They all came on the assumption that rather than fighting over who had the largest share of some pie, it was better to collaborate to make the pie a whole lot bigger.

At TPAC, disputes could sometimes grow heated. Often, technical disagreements had been simmering in email listservs and discord channels for months. I felt it was vitally important that the arguing parties be given a chance to meet face-to-face to discuss their grievances and come to some sort of consensus. Just as important, the minutes of every meeting, no matter how trivial, no matter how contentious, were posted online, so that there would always be a record of how decisions were reached.

I think of the TPAC summits as an important global force for technology. I know of no other organization in the world that brings together such a diverse group of corporate, government, academic and non-profit technology interests. Negotiations can get a bit chippy, and sometimes progress is slow, but the end result is standards that foster seamless interoperability across languages, cultural barriers, platforms and technologies. TPAC really put the 'World Wide' in the World Wide Web. As proud as I am of the work I did in designing the original protocols for the web, these annual summits have been crucial for keeping it on track. As I write this book, though, I ask myself how we are doing with the latest wave of tech. It's a question I'll return to.

•

The World Wide Web's growing popularity offered me the chance to meet and get to know some pretty interesting people. While I'm not exactly a celebrity – I don't get recognized on the street – I did receive a number of awards. In 2004, I was named the inaugural winner of Finland's Millennium Technology Prize. Finland had conceived of this award after noticing that the absence of a Nobel Prize for technology meant a large number of important innovators and inventors weren't being recognized. Since Norway awarded the Nobel Peace Prize, and Sweden the other Nobel Prizes, I suspect Finland also saw an opportunity to join that group. The web was selected as the winning technology from among seventy-eight nominations from twenty-two countries.

This award meant a great deal to me personally; I was very honoured to receive it. I travelled with my family to the Finnish National Opera House in Helsinki. Tarja Halonen, the president of Finland, presented me with the award. The members of the symphony, and everyone else in the audience, stood to applaud.

The physical prize was a sculpture made from crystalline silicon; it was accompanied by a cheque for €1 million, which I certainly appreciated. But the real reason the Millennium Prize meant so much to me, personally, was that I'd brought the family, including my children, Alice and Ben, along for the trip. They were now aged thirteen and ten, and it was the first time we had done something open and festive as a family involving the web. We had always kept the web thing low profile at home and at their schools. Alice and Ben had a great time in Helsinki, a charming city on the Baltic Sea with beautiful waterfront views and numerous islands to explore. We stayed in the British Embassy, courtesy of ambassador Matthew Kirk and his wife, Anna, who were wonderful hosts and helped explain Finland a bit. Their red Jaguar, festooned in a Union Jack, transported us around town (to Ben's delight) – and the whole embassy was a safe retreat for us to regroup after busy and demanding days.

The Finnish visit was one of the last official trips I took with Nancy. Our marriage in fact came to an end over the next few years, and I was hopeful for new horizons in my work and personal life.

•

Even as the web was extended to ever more people, its integrity was threatened by Microsoft's growing platform dominance. By 2002, Netscape was in shambles, and Internet Explorer had a 95 per cent market share. To outsiders, the browser wars looked to be over. Around this time, I drafted a map of the web, inspired by J. R. R. Tolkien's map of Middle Earth. The happy stream led from the shire of the World Wide Web to the tranquil Sea of Interoperability. To the south, however, surrounded by the Wasted Arid Plains, the Patent Peaks and Proprietary Pass, loomed the sinister, anagrammatic Tor of Cism. If you zoomed in, you would see that the eye of Sauron is in fact the 'e' on the Internet Explorer logo. The map still hangs on the door to my MIT office today.

Fortunately, Microsoft's browser monopoly proved impermanent. The non-profit Mozilla foundation, formed from Netscape's ashes, launched the alternative browser Firefox in 2004. Firefox did a lot of things well, and we pushed hard to make sure Mozilla had a seat at the table at W3C. It was also the browser I used most of the time, although I also occasionally experimented with Opera, an alternative browser from Norway. In the mid-2000s, Firefox began to eat away at Microsoft's position, and eventually Internet Explorer lost its monopoly position. Explorer remained an important player for many years, balancing Mozilla, Opera, Safari and later Chrome. (Today, its successor, Edge, has only about 5 per cent market share.)

But even as one monopoly disintegrated, another began to form. By the end of 2004, Google had eclipsed Yahoo as the dominant search engine and would eventually grow to control about 90 per cent of the market for internet search. I found a lot to admire in Google. They were a very innovative company that did a great many things well, and they were committed, for the most part, to maintaining open, royalty-free standards. They were especially good at using the programming paradigm known as 'Asynchronous JavaScript and XML', or the AJAX platform, which allowed web data to be retrieved in the background without interfering with the behaviour of the existing page. This meant users could interact with rich, feature-filled applications without constantly hitting the browser's refresh button. Google used AJAX to develop the popular Gmail application in 2004, and Google Maps in 2005. These beautiful, innovative products created the illusion that the web was always 'on'.

How do you make an application that seems to be 'always on'? Two ways. You actually keep a copy of much of the web – especially maps but also other data – on the device so that what you want is available instantly, even if the connection is slow or has dropped. Then, if the user makes a change to what they're looking at, say moves something to the map, you make the change locally on the device, and then forward the change to the system, the back-end store or elsewhere on the web.*

Due to its predominance in search and its innovative products, Google soon had a lot of clout inside W3C, and Google developer Ian Hickson was named the chair of the HTML development group. Hickson, or 'Hixie' as everyone called him, was a Swiss national who had grown up in Britain. He was technically capable and highly intelligent, but rather combative. He and his cohort didn't much care for W3C's consensus-building approach, which they found boring and slow. Hixie especially didn't seem to like the horizontal review by the accessibility community.

Hixie's grumbling led to problems as the HTML standard evolved, but his role was important. As an employee of a search engine company (Google), he was seen as a neutral arbiter between the fiercely competing browser manufacturers, namely Microsoft, Mozilla, Apple and Opera. But soon, two factions broke out in W3C, with a rather toxic culture based on confrontation vying against the older norm of engineers collaborating together respectfully to build the best world they can for everyone.

The debate was over what direction the fifth version of the HTML standard should go. Should HTML now be encouraged to become cleaner and simpler? The problem was that the actual web

* In 2025, doing this with multiple collaborating users is called 'local first'; it is an important little movement among developers.

pages that real people had built were not at all clean. They had hundreds of errors, where users or developers had made typos, or misunderstood the HTML protocol, or quite commonly, had simply copied the mistakes of another person's web page to their own.

This was due in part to the unique way the web had spread. Many people, rather than building their own web pages from scratch, would instead navigate to a page they liked and select the 'View Source' option in the browser. They would then copy the source code, modify it with their own stuff, and upload it to a server. This helped the web to grow quickly, but it also caused (sometimes quite serious) programming errors to propagate across websites.

Now suppose, instead, the browser said, 'Here is your friend's web page, but cleaned up a bit' – then the same typos would not be repeated! That's what *I* wanted, but the browser companies didn't. They wanted to pass on the old web page, warts and all. In fact, they did worse. During that time, it became a matter of pride in each browser to be able to display absolutely *anything* that anyone put out there on the web. They were competing in a race to the bottom.

Hixie's browser group wanted browsers to read every page, even pages that were poorly coded, that didn't meet basic data standards, and that were riddled with errors. Their spec defined how to read a web page in English, step by step, which was very tedious to have to code. The stated motivation was to hide all this dysfunction from web users, who often didn't know (or care) about how a web page was designed. Of course, doing it this way shifted most of the work to the browser application, which didn't seem to be a coincidence.

Indeed, you could argue that Hixie's coalition of Safari, Opera and Mozilla were protecting their little clique, by making it harder for new browser developers to enter the field.

Following a period of intense debate, in June 2004 the matter was put to a W3C vote, and my preferred, 'cleaner' approach won. I figured that was the end of that, but what happened next came as a disappointing surprise: representatives from Mozilla, Opera and Apple broke away from W3C to create a splinter faction. It was called the Web Hypertext Application Technology Working Group or WHATWG. (My friend Hakon Lie defected as well, which was disappointing too.) WHATWG formed their own mailing list and continued developing their flavour of HTML beyond W3C's control. There was a real toxicity to the WHATWG mailing list; they were a bunch of typically white male geeks who had hoisted the pirate's black flag, and they were not very polite about disagreement.

WHATWG's approach allowed web developers to make all kinds of errors, and never complained to anyone reading the page by raising alert boxes or banners. Now, because the errors in the web page are hidden from the user, they are also hidden from the *developer*, who can never see what needs to be done to clear up the messy bits. Under this approach, the web would become bloated and slow, full of errors which the browser was programmed to ignore.

Hixie soon left W3C completely, and not long after, someone found a series of blog posts he had previously written advocating for what he called 'humanitarian eugenics'. Hixie's callow manifesto called for selective reproduction based on measures of intelligence, where mothers cannot bear offspring without a licence. As you can imagine, this did not go over very well within W3C. But it turned out he had another surprise up his sleeve too.

It was soon revealed that Google was developing its own browser, known as Chrome. This was very smooth of Hixie, and nobody saw it coming. By getting editorial control of HTML, he had been able quietly to make sure that Google had what it needed for its own secret project. Soon, Mozilla, Opera and Apple realized they were working not with a neutral convenor but a rival, who had been steering HTML in the direction that Google's Chrome team wanted it to go.

In my time at the head of W3C, I witnessed a lot of slick bureaucratic manoeuvring by various special-interest groups. I would say Hixie's hijacking of the HTML standard was probably the slickest. Google Chrome publicly launched in 2008, and it was an immediate sensation, its success due in part to its seamless integration with the HTML5 standard WHATWG had produced.

Now, I want to give credit where it's due: even though it propagated a lot of errors, HTML5 was a more powerful evolution of web technology, and we ultimately endorsed it within W3C. It essentially turned HTML into a programming language for JavaScript developers and eliminated the need for third-party plugins like Flash and Shockwave. So-called 'Rich Internet Applications' like Google Docs and Slack began to appear, consolidating almost all of the functions of the PC inside the browser. Google Chrome is in 2025 by far the most popular browser in the world, and it's one of the ones I use most often. It's a fantastic product, no question, and the modern, browser-first paradigm simply wouldn't be possible without HTML5.

But: the fracturing of W3C left the world worse as a whole. The formation of WHATWG was, to my mind, the first real blow to the integrity of the World Wide Web. It made the web less accountable to public stakeholders, and more of a bloated, complicated mess. The schism was, to me, the first step away from the public commons I had envisioned, and the first step towards something far more divisive and exploitative. Unfortunately, it would not be the last.

CHAPTER 9

The Mobile Web

As the web conquered the PC, we did our best to keep it open and to help it improve. Meanwhile, however, we faced an even bigger challenge: the looming adoption of mobile devices. Today, that adoption is complete: people sometimes spoke of the 'transition' from PC to mobile, but there are a huge number of people in the world – probably a majority – who have *only* ever used the web on mobile phones.

At W3C, we saw this revolution coming; we started our Mobile Web Initiative in 2005, more than two years before the introduction of the iPhone. The initiative was led by Philipp Hoschka, a German computer engineer with a PhD from the University of Karlsruhe who recognized quite early that the web would someday be bigger on mobile than it had ever been on its native environment of the PC. To prepare for this eventuality, we had to change the way people made their websites.

The first and most important change was the introduction of 'haptic' touch events, which took the place of clicking on links with a mouse. Resizing the links to correspond to the press of a finger to a touchscreen was something of a project. While I had conceived of a wide variety of ways to *embed* links in the web – in text, via pictures, in interactive objects, etc. – my default expectation was always that the user would have a mouse and keyboard in front of them. Touchscreens were a new paradigm, and one W3C struggled a bit with at first. Fortunately, Apple was one of our partners at W3C, and they pushed Hoschka and his team to get this working fast. (In 2007, with the launch of the iPhone, we found out why.) The diminished screen size of mobiles also presented new challenges. Fortunately, the separation Hakon Lie and I had engineered between style and content many years before made it relatively easy to resize web content. The other challenge was finding a balance between phones' landscape and portrait modes. Apple helped us here, as well.

Over the long term, probably the most important change mobile engendered was the shift in users' attention spans. This was more of a content problem than a technical one, but *everything* had to be simplified – for web designers, ample white space and ease of rendering took precedence over the elaborate arrangements of the websites of the early 2000s. The shift to mobile necessitated a refocus of ambition in graphic design. The upside was ubiquitous connectivity.

It's very important that this transition happened. For the web to be ubiquitous, it had to be available on phones. And to make the web affordable – to get the data rates down so that it was accessible to less well-off users – we had to focus very much on the smallest, least powerful device. The Mobile Web Initiative realized this was important, and fortunately we could address these constraints via the power of the cascading style sheets we'd devised earlier. With CSS, you can assign different parts of a style sheet for rendering on a mobile, or laptop, or a large screen. Having different ways to serve the same content had always been key for the web: for accessibility, for internationalization, for whatever device, the web always had to be universal.

Designing for mobile meant guiding developers towards a new type of content presentation. The mobile phones supported all sorts of input and output channels that laptops of the time did not. These included touch events, vibrational feedback, accelerometer data and, of course, current location data. The last was very tricky, as we had to make sure websites could take full advantage of knowing where the user was located while still respecting privacy concerns. And there were other possibilities we wanted to take advantage of, even in the earliest days of the smartphone – like mobile videoconferencing. Impressed by the early Nokia Communicators, we had been working on that well before the debut of the iPhone.

The transition was smoother than I expected. Perhaps it was because, with all the work on accessibility, W3C folk were well versed in thinking about different situations in which the same information needed to be rendered in different ways. The W3C mobile standards were up and running by the time the smartphones debuted, spurring a virtuous cycle of development and adoption.

•

In July 2004, I was made a Knight Commander of the Order of the British Empire, in recognition of 'services to the global development of the Internet'. From that point forward, my official designation became Sir Tim Berners-Lee (you never say Sir Berners-Lee, it's Sir Tim on special occasions; but call me Tim). Getting knighted at Buckingham Palace was a special moment in my life. When we met, Her Majesty had someone standing behind her, whispering who each person was. As it came to my turn, she asked me how long I had been involved with the internet – a reasonable question – though I stammered, nerves taking over, and couldn't produce a coherent answer. Her Majesty was a great conversationalist, however, and the talk continued. As she touched my shoulders with the knighting sword, I felt proud of the movement I represented and the crazy community of collaborators all over the world who were helping to make the web better. There is no big party at Buckingham Palace after the ceremony, so we celebrated by going out to lunch at the Savoy.

The knighthood was of course a very great honour for me, one I shared with about 3,000 other living knights and dames of the British Commonwealth. I was further honoured to receive what came later in 2007, when I was appointed to the Order of Merit of the British Empire. Only twenty-four individuals are invited into the Order of Merit at any given time, and it is our privilege to meet for lunch with the monarch every few years. (We get to bring our partners as well.) The Order typically convenes at either Buckingham Palace or Windsor Castle. Attendees are expected to wear their honours, which the Royal Family's website describes:

> The badge is an 8-pointed cross of red and blue enamel surmounted by the Imperial Crown. In the centre, upon blue enamel and surrounded by a laurel wreath, are the words 'FOR MERIT' in letters of gold.

The food at the luncheons is delicious, although you do have to know how to tackle it sometimes. There are a lot of knives and forks – the key is working from the outside in. Of course, some of the finer points of etiquette can only be arrived at through experience. Artichoke, for example, is eaten with the fingers; afterwards, you rinse off your fingers in the beautiful brass bowl sitting next to your plate. Fortunately, while working at D.G. Nash in the late 1970s, I'd once dined on artichoke with some officers in the French Navy, so I knew the drill.

At my first lunch with Her Majesty, I found myself seated among some very interesting people. The neat thing about the Order of Merit is that its members are leaders in totally different fields, some of the greatest living Britons. David Attenborough, the famous BBC nature presenter, was there. The painter Lucien Freud, grandson of Sigmund, was in attendance as well; he was rather quirky and would turn his face sideways in the class photo that marked each meeting and was published in the press. Let's see, who else? There are too many amazing people to mention here. Roger Penrose, the brilliant physicist who had been the idol of many of my fellow students at Oxford, and also my friend from a biking group, Norman Foster, were a delight to talk with. And once, following the first OM lunch, we shared a cab home with the playwright Tom Stoppard, who stopped at a little higgledy-piggledy bookshop by the side of the road to purchase for me a copy of Alan Bennett's *The Uncommon Reader*, a fanciful novella which imagines the Queen abdicating from the throne to devote more time to pleasure reading.

The star of all the attendees was of course Queen Elizabeth II herself. To my great surprise, Her Majesty seemed to enjoy my presence at these functions, and over the years I was tremendously pleased to get to know her a little. She was a great inspiration to me, as I suppose she was to people all over the world. I often felt that we had a mutual bond and respect for public service. In 2009, we helped her roll out the official website for Buckingham Palace – the Queen launched the site with the push of a button on a remote control. 'I think we did that rather well, didn't we,' she said to me later. I wondered, was that the royal 'we', or did she mean both of us?

We were lucky to be able to experience Her Majesty's wonderful gentle wit. My wife Rosemary, a Canadian, also enjoyed our interactions with Jean Chrétien, the former prime minister of Canada, and another member of the Order. When Jean realized both my wife and my Granny Helen were Canadian, he remarked 'The web is Canadian! Canada can claim credit for one of its own inventing the web.'

•

I first met Rosemary Leith at a dinner in 2008. The hosts were my friends Nigel Shadbolt and Wendy Hall, two professors at the University of Southampton. Nigel, Wendy and I were all on the board of the Web Science Trust, a charitable trust which supports interdisciplinary research into the effects the web has on society at large. They were courting Rosemary to join our board.

Rosemary was a highly accomplished businesswoman with an enormous gift in her ability to connect with people and understand how they think. She holds others and herself to a very high standard. She had grown up in Canada, then moved to the UK via Switzerland in the late 1980s to work for the private equity firm Pallas. In the late 1990s, she co-founded an early content company called Flametree, a dot-com that took advantage of web technology to provide flexible work–life solutions. Rosemary and her co-founder were very far ahead of their time; both had relied on flex-work in order to succeed in business while simultaneously raising children. Flametree was a success, and in 2001 they sold it to PwC just three years after inception. Rosemary then worked as a venture investor, using her ability to recognize early shifts in society that impacted technology and to see their resulting commercial impact.

I found in Rosemary a strong-willed, highly intelligent woman. She had dark hair, dark eyes and a winning smile. I liked her right away, and I began inventing reasons to see her again. I had a TED talk coming up and, although I didn't know her very well, I asked her to help me write it. I believe she was a bit sceptical of me at first – but why don't I let her tell this part of the story?

'I almost didn't go to the dinner that night,' Rosemary recalls. 'Tim and I existed in very different worlds that had very little if any overlap. Had I not gone to the dinner we would have never met, a sliding doors moment. As I got to know Tim, I realized how similar our upbringing had been, even though we were continents apart.

We shared a great love of the outdoors, but I needed to test him, to see if his outdoor skills were genuine. So I took him on a several-day canoe trip to see if he could survive in the Canadian wilds. We camped at night and the first day we paddled all day and arrived at the area we had planned to stay. We found a great camping spot and wonderful rocks from where to swim. Tim was competent with the canoe and relaxed into paddling. He could pitch the tent, build the fire from nothing, and bear-proof the campsite! That totally impressed me.'

'I found in Tim such a mix,' Rosemary continues. 'On one hand, deeply technical but with a love of music and the arts – a painter. He was also a student of history and philosophy, coupled with an openness to others' ideas, complete respect for humans and nature and entirely comfortable with strong women. I could see how it came through his mother, with the magical childhood that he had. And while he was obviously brilliant, he was also methodically adventurous and had a warmth with people. He liked to hike, and bike, and swim. He was a little dashing, actually, once you got him in the right element.'

'He's a special person, beyond giving away the web,' Rosemary says. 'He's open-minded, fair, resourceful and very kind. But what I think really sets him apart is that he cares about people as much as tech. He truly cares about those less fortunate. He loves to walk and be outside; he can be happy with the smallest amount of material things. Also, he picks up garbage everywhere he goes. He'll walk with a garbage bag on a hike, constantly stopping. Good for the planet, but makes for a very slow-paced walk.'

With Rosemary's help, the TED talk was a success. By coincidence, Bono, the lead singer of U2, was giving a TED talk around the same time. We met each other in California and got along well. Bono invited us to watch a U2 concert some time later, and when we went he changed the lyrics to his song 'Beautiful Day': 'And there's Tim Berners-Lee, standing right in front of me.' Wow. We haven't seen Bono again so far, but it was a pretty good date, I must say. We both ended up with massive respect for the man, the music, and the mission.

Rosemary and I began to travel together often. Through my speaking engagements and my work with W3C, I was already moving around the world quite a bit, but once I met Rosemary my life became an almost non-stop flurry of activity. 'We're both high-energy people,' Rosemary says. 'We have a pretty hectic schedule. We're both happy that way.'

Rosemary and I were co-founders of the Web Foundation. Rosemary shared my concerns about the widening global access gap that the web was creating between the wealthy and the underprivileged, who often lacked access. The web, we observed, was getting stronger, but not more widely distributed, and we thought we should take action. The idea for the Web Foundation was really both of ours, and our responsibilities split pretty naturally. 'We have complementary skills,' Rosemary says. 'Tim's an optimist. I provide some scepticism and challenge – I'm more of an entrepreneur.'

In W3C we worked hard to make the web better, but we were aware of a wide range of ways in which the web had a negative or dangerous effect on the world. I felt a strong sense of obligation to address these serious concerns about the web – while also increasing access to it. Every time the people in W3C made the web more powerful they, in a way, increased the gap between the haves and have-nots. The digital divide got greater. At the same time, there were social impacts that were not being addressed – from political polarization to online gender-based violence.

It made sense for me to initiate some fundraising to help tackle these issues; Rosemary handled strategy and building the management team. My initial fundraising efforts weren't so successful, until I had a chance encounter with Alberto Ibargüen, then the president of the non-profit American Knight Foundation, at his flagship event in Washington, DC. We were queuing up for food at the buffet together, and I figured I might pick his brain about the best way to start.

Now there's an old saying about the non-profit world that I have found to be sometimes true: 'If you ask for money, they give you advice, and if you ask for advice, they give you money.' And as I asked Ibargüen about what to do, he confessed that the looming digital divide was a matter of great concern at the Knight Foundation, and that he'd been looking to fund exactly such a project. 'I can write you a cheque for $200,000,' he said, and told me he would pitch his idea to the board that same day.

Sometime later, after we completed the grant applications and went through a detailed diligence process, Alberto agreed to support the launch of our Web Foundation. He had the authority to grant the $200,000 we asked for, and he invited us to launch in Washington at the Newseum which the journalist Shelby

Coffey and the Knight Foundation had just finished building. We arrived in Washington and were in the hotel, dressing for dinner, when the phone rang. It was Ibargüen: he told us the board had authorized him to give us $1 million. I looked at the phone, stunned, then stammered out my thanks. 'No, Tim, you don't understand,' Alberto said. 'The board felt that, while a million was a good number, a year was too short. We are giving you $1 million a year – for the next five years.' So that was a good trip to DC!

With Alberto's assistance, Rosemary and I were able to get the Web Foundation started quickly. Rosemary's commercial experience helped us bring in some influential figures who put us right on the map. The UK prime minister, Gordon Brown, and leading businesswoman Helen Alexander joined the board. Then Alberto himself agreed to chair it. They were all supportive of the mission but I suspect that, without Rosemary's ability to engage, articulate and cajole, we might have been a party of two. We launched in November 2009. The first CEO was Steve Bratt, who'd been managing W3C's industry consortium. With support from other donors, we soon had an operating budget of $5 million and fifteen people working in the team.

The aim, as I say, was to make web access a basic human right, similar to food, medicine and shelter. The most immediate task was to increase the web's global presence. For Western users, the internet was by 2009 a well-entrenched technology, but only 20 per cent of the world's population had ever logged on! We set a target goal of getting that to 60 per cent by 2019 – the modest objective of getting 3 billion more people online.

Rosemary took the lead in setting out the foundation's strategy. Looking at where we could get the most leverage for our funding for the most impact, the board concluded that the best investment we could make was in policy efforts to increase access in the developing world. We felt, though, that we really should investigate the state of internet connectivity and use, so we should go there ourselves and find out.

Thankfully, the BBC asked whether I would go with a reporting team to Ghana to see the state of the internet there. Aleks Krotoski, their journalist, was the lead on an expedition to see whether the stories of abundant internet cafes and farmers online were true. It was fun driving around the jungle in the BBC's little blue Land Rover. We talked to young people in villages who had good English and online friends in the USA. I encouraged them to blog about their own world and to put their own town on openstreetmap.org. We noticed that there was mobile data signal at the top of each hill, even in rural areas. But we also saw that smartphones and data plans were too expensive for most: an issue the Web Foundation would later address.

Later that year, we travelled to Uganda, where the government put on a very big show of it. Rosemary and I were met at the airport by a large security contingent, then we were escorted into two limousines, and driven in a high-speed motorcade to government headquarters, running other cars off the road. I travelled with a minister, who told me the biggest problem Uganda was facing was its population explosion, with a huge number of children under the age of ten. 'How are we going to feed and educate all these people?' the minister asked.

I couldn't help myself. 'How many children do you have?' I inquired.

'I have twenty-three,' the minister said.

We had a day or two of meetings with governments and organizations, all with official drivers and security, and then we planned to drive across the country to see the rural areas. The next morning, though, our motorcade and security detail mysteriously failed to arrive. As we had a number of other meetings scheduled throughout the country, Stephan Boyera, our trusted Web Foundation colleague, ended up having to hire a driver in a beaten-up Volkswagen van to get us across the country, including a magical stop at the equator. At one point the vehicle broke down and we had to stop to find someone to repair the van, going to a nearby farm to get water.

It turned out to be a preferable way to tour the country, as we could see what was really going on, rather than what the government wanted us to see. Uganda, at the time, was not yet well connected; I recall stopping at a small restaurant with a sign reading 'Internet Cafe' in front. This cafe did not, in fact, have internet – it was just the name of the place. Later, we passed a troop of workers walking past carrying bundles of cable. Under the supervision of a foreman, they were digging a trench and putting down broadband. We stopped the van to talk to them, and, with the assistance of the driver, learned the men were being paid one dollar a day to build a fibre optic system that would stretch from the port of Mombasa all the way to the Democratic Republic of the Congo. One way or another, the world was coming online.

Just after the van broke down! Excited to be on the equator.

Further visits to different parts of Africa helped us to understand the state of web connectivity and how we could build projects that would increase web access and its meaningful use. Back in late 2008, at an MIT president's Christmas party, we had the good fortune to meet MIT alumnus Jono Goldstein and his wife, Kaia Miller. Both Jono and Kaia had been active in Africa – and Kaia specifically in Rwanda. Jono, an MIT engineering graduate, was organizing a trip to Africa in 2010 and we decided to join them and build a journey around the visit. This gave us a good reason to embark on a tour of East Africa, meeting with government officials to talk further about increasing affordable web access.

First, Rosemary and I travelled to Nairobi, where we were hosted by the British High Commissioner. (A high commission is, for a Commonwealth country, what is normally called the embassy.) We arrived to find the wireless transformation already well underway. Talking to average Kenyans, I would always ask what they looked for online. Many used the web for football scores; others were using it to surf the web and find out more about the White House! Above all, even then, everyone was obsessed with Facebook.

After a couple of days of meetings with Kenyan government officials, it became clear that the key thing was making sure that internet services didn't end up controlled by a monopoly. There were a lot of vendors vying to be the sole provider of broadband in these countries. They'd offer the governments a Faustian bargain: the gleaming prize of free wifi in all of your schools, in return for monopoly access to all of your population. For ministers in impoverished countries, it was a hard offer to refuse.

Part of our work was simply encouraging these governments to hold out for more. Once you did, other providers moved in quickly with competing offers. You could truly discern the health of the marketplace in an African country by looking at the colour of the neighbourhoods. The mobile providers were paying people to paint their houses – red for Vodafone, say, or yellow for MTN (it was less ugly than a billboard, I suppose). It was healthy when there were a lot of different colours.

One of the leaders in regional connectivity was Rwanda. President Paul Kagame was trying to transform this small country into the Singapore of Africa. He already had his entire palace wired for internet access, with nine international gateways planned and routers in place. After his forces stopped the genocide in 1994, Kagame had imposed order in Rwanda – sometimes brutally, and brooking little dissent. His seizing control of the radio station in 1994 was the key to stopping the genocide, and he had subsequently banned all ethnic identification. Around the country, you could drive past banners strung up on trees that read, 'No Genocidal Thoughts'. On the trip, we went to see a girls' school in a rural area. When we got there, they were having trouble with the internet. They had this huge, clunky old satellite, but the connection was down. When I arrived, they asked, since I was the inventor of the web, if I could fix it. Sometimes people do ask me to fix their personal internet connection.

These trips were such an amazing opportunity to see first-hand both the power of the web for change, but also where the need was most. They also gave us time to build some incredible relationships of strength. For instance, Jono Goldstein later became one of our longest-serving board members with an uncanny ability to understand both the technology and how it might benefit the developing world. Our work benefited significantly from his expertise.

•

The rapid global adoption of smartphones led to an explosion in the number of people and devices online. Vint Cerf and the IETF had to expand the IP addressing system by several orders of magnitude just to accommodate all these new devices. It had taken the web about twelve or thirteen years to get to 1 billion users. Two years after the introduction of smartphones, we reached 2 billion.

W3C opened its India office in 2010 and we then hosted the web conference in Hyderabad in 2011. During that trip, I was invited to discuss an ongoing programme to put web technology in the hands of the country's citizens. We were given an audience with Nandan Nilekani, co-founder of the tech consultancy Infosys, and we asked him about the push to get government data on the web. In answer, he turned to the overflowing filing cabinet sitting next to him and gave it a tremendous whack. 'It's not on computers!' he said. 'Before we can put it online, we have to get it onto a computer.' Nilekani was looking to replace the physical paperwork with Aadhaar, a program that would give everyone in the country a trackable, biometric digital ID. Aadhaar was quite remarkable. Streamlining the enormous machinations of a bureaucracy overseeing 1.3 billion people posed countless challenges, and the system wasn't always perfect, but replacing those filing cabinets with databases connected to the internet was a huge step forward.

We've returned to India a few times since then and have gradually watched the country change since that day. Today, India is online and functional in a far more significant way than on my first visit.

We had a contrasting experience visiting South Korea. Even in the early 2010s, smartphones were already ubiquitous there, and the wireless data infrastructure was some of the best in the world. In fact, I was astonished by the place: the people were friendly, the food was delicious, and the climate was similar to Europe, but the culture was radically different. Several government ministers solicited my opinion on various technical matters during my visit, but the truth was that there was little I could offer by way of advice – in many ways, South Korea seemed more technologically advanced than the West.

Around this time, the Web Foundation ran a project in Burkina Faso, in West Africa. The difference between the plugged-in citizens of Seoul and the subsistence farmers of the Sahel was huge.

The Sahel is the transition zone on the border of the Sahara; it is a dry, difficult place to live. A long drought in the 1980s had once threatened to turn it into a dust bowl, and the rains, when they do come, arrive all at once, as flash floods during the brief wet season. In this barren environment a Burkinabe farmer named Yacouba Sawadogo was using an ingenious planting strategy. First, Sawadogo made a hole in the ground with a stick, about 6 inches across and 6 inches deep. He then fertilized the hole with animal dung and planted a seed. When the floods came, the water would collect in the hole rather than washing away, creating a kind of makeshift hydroponic system. The manure also attracted termites, who acted as insectile cultivators, breaking up the hard soil.

The technique, called Zaï, had ancient roots, but Yacouba had revived it for the modern era. In the 2000s he took on an unforgiving patch of arid land south of the capital city of Ouagadougou and, using Zaï, transformed it into a vibrant, 25-hectare forest.

Now what does this have to do with the web? Well, by teaching this technique to other farmers, you could grow crops in arid zones that previously looked unfarmable. The key was connectivity. With funding from Western NGOs – non-governmental organizations – Sawadogo was given a smartphone to promote the technique, and he and other Zaï experts created informational videos in a variety of local languages. They also acted as remote consultants to other farmers in the region.

The Zaï technique could have spread without the web, but it might have taken decades. With the web it happened instantly. Once Sawadogo was given a platform, Zaï farming spread throughout Burkina Faso and into the neighbouring countries of Mali and Niger. Ultimately, about a half a million hectares of land were cultivated, and farmers planted not just trees but cereal crops like sorghum and millet. The work was not easy; Zaï techniques required labourers to dig hundreds of holes in densely packed soil in the heat of the equatorial sun, typically without cover of shade. But in the end the reclaimed land provided enough food to nourish an estimated 3 million people.

Although the web spread wealth throughout the world, our experience in Burkina Faso suggested that its most transformative impact might be in Africa. With the rise of the mobile smartphone, people who previously had no connectivity whatsoever – neither computers nor televisions, nor even landline telephones – might experience rapid economic growth and industrialization. Helping to connect Africa and the rest of the world to the web seemed to me one of the most valuable uses of my time, a way to affect a great number of people directly. The idea was to spread the benefits of the web to even the poorest regions on the planet – or, as the Web Foundation's mission statement read, 'to advance the open web as a public good and a basic right.'

•

The campaigns for fair internet access were not limited to sub-Saharan Africa. In fact, I often found myself fighting them in the United States. In 2007, I was asked by Congressman Ed Markey, a Democrat from

Massachusetts, to testify before the US House of Representatives on the contentious subject of net neutrality. I had known Markey for quite some time. He was a far-sighted legislator, one of the few, alongside Al Gore, who actually took the time to understand what the internet was. From 1987 to 2005, Markey had chaired the congressional telecommunications subcommittee, a job that required him to stay abreast of the latest developments in both mobile and internet technology. He took this position seriously and spent a lot of time talking to teaching staff at MIT about how to bring about the telecom revolution. In 1993, Markey had introduced legislation that auctioned off the US mobile phone spectrum, creating licences for seven new providers and opening up a competitive marketplace for cellular service. Prices for mobile phone service plummeted, and soon everyone was going wireless. That was what good policy work could do.

The next regulatory frontier was the cable companies, who ran the broadband data pipes into Americans' homes. Cable was a sort of natural monopoly, like electricity, water or sewage; it was impractical to build competing services for the same neighbourhood. But the cable companies weren't as tightly regulated as most municipal utilities, and they used their monopolistic position to charge high rates for mediocre service.

Because of the way the internet addressing system had been built, the cable companies could see who was sending which data packets where. This allowed them, at least in theory, to charge for preferential treatment of certain customers' data packets. For example, the cable company might charge Netflix a premium to get its data down the pipe first, delaying the delivery of emails and other 'non-essential' services.

In the past, the cable companies had always controlled what got into your house, and what got onto your TV. They did this by creating bundles of content at different prices, then blocking access to anything which was not in the bundle you had paid for. You can imagine then how running the internet over the cable and allowing people to access whatever they wanted on the web went against the grain!

The principle that every data packet on the internet be treated equally was known as net neutrality. Net neutrality means that when you get your internet connection you can use it for whatever you like. It means the market for content and the market for connectivity are quite independent – and with that policy in place, both markets have flourished over the past thirty years, to an almost unbelievable degree. Without net neutrality, the web assumes a different form: it would be as if when you bought a car of a given make, you were also buying a package that included what roads you could travel on and the destinations you could go to. Fords might have to travel on certain roads and Volkswagens on others.

Think about it: when the web started, modems were 300 bits per second – those acoustic couplers on your handset. Now, broadband connections are 30 megabits per second – that is 100,000 times as fast. Nothing in the history of anything goes 100,000 times as fast. And meanwhile, the web has spread from one website to include all the crazy stuff we have now. But people developing the content never worried about what net speed would be available, and the people working on network speed didn't in general worry about developing new content for the web. Their work proceeded independently, because of net neutrality.

The free, open access communications paradigm we have did not arrive like magic. It was the product of a fair amount of political wrangling – we had to convince governments worldwide to deregulate the telecom industry; we had to fight for the 'safe harbour' provisions to ensure web hosts weren't liable for offensive or defamatory user content; and we had to ensure a fair auction of the radio spectrum for wireless communications. To many, protecting the principle of net neutrality – the equal flow of traffic across the data pipes – was merely the latest stage in a long political fight.

At the time I testified, the Democrats were in control of Congress. While I try to maintain a politically neutral view when it comes to the web, I can tell you that Ed Markey and Al Gore took the time to understand the technology better than their counterparts on the opposite side of the aisle. Apart from Markey, at that time it felt as if nobody on the committee really understood what the internet was or how it worked. That said, it was good that a bunch of politicians would find the time to consider the relationship between technology and society.

CHAPTER 10

This Is for Everyone

Back in the days when Facebook was one of the many competing social network sites, I found it was pretty useful. It reconnected people with old friends and classmates, and allowed people to make new social connections. In the best scenarios, it could even accelerate your offline life. In those early days, social media seemed like a positive step in the evolution of the web.

Facebook had its conceptual roots in other, earlier web communities – sites like MySpace and Friendster, or before that LiveJournal and even forgotten services like AOL Hometown. The problem with those services was that they owned the data the user generated. This wasn't how I had envisioned the web working – in my vision, users had complete control *and ownership* of their content. In fact, when AOL Hometown went kaput, they basically deleted everything their users had built. This happened later with Friendster and MySpace as well, eliminating the value of many millions of hours of creative work that users had put into their pages.

I first met Mark Zuckerberg in the early 2010s. He was fairly young back then and still liked to wear his hoodie. We were scheduled to meet at Facebook's offices, but he had come down with the flu a few days earlier. Instead, I travelled to his home, where we sat across the table from one another and ate chicken noodle soup. I was impressed by Mark; he was obviously very bright and ambitious. And I was very grateful when Mark subsequently directed Facebook to make a donation to the Web Foundation.

If I was impressed with Facebook, then I also was equally impressed with Twitter, too, when I first saw it, and Instagram as well. People sometimes grouped these services under the term 'web 2.0', implying a new emphasis on sharable content, but the label meant little to me. The entire purpose of the web, from my earliest conception of it, was a place to share new user-generated content. What social media did well was to organize that information into easily parsable channels. Unfortunately, what ended up happening was that, as the technology evolved, a handful of providers grew into dominant, unregulated monopolies: YouTube for video, Instagram for images, Twitter for microblogging, and Facebook for one's personal launchpad. These digital platforms began to replace the older, more organic experience of building one's own platform using the HTML tools of the web.

I had mixed feelings about this, as you might imagine – and gradually, I began to grow disillusioned with Facebook. With the help of his COO Sheryl Sandberg, Mark monetized his platform by bundling Facebook users' data and targeting them with advertising. By matching Facebook data with search histories, advertisers could build detailed demographics of users, then follow them all over the web with ads. I felt this was a privacy violation, and in talks, I began to speak of the concept of 'data sovereignty' – the idea that *you*, not Facebook, should own your profile and your history of interactions.

•

The first large-scale demonstration that social media was going to transform society was the 'Arab Spring' protests. The wave of unrest began with Mohamed Bouazizi, a street cart vendor from Tunisia who had suffered a lifetime of harassment and petty extortion from regional officials. The final straw came on 17 December 2010, when his digital scale was confiscated by a local police officer. Bouazizi went to the governor's office to complain; when the governor refused to see him, Bouazizi stood in the middle of the road outside the building, doused himself in gasoline and set himself on fire. He died a short time later.

Reports of Bouazizi's actions spread rapidly on Twitter and Facebook. Within a week he'd become a symbol for resistance against the ruling Tunisian regime, a corrupt and repressive one-party state that had continuously held power for years. Leveraging the network effects of the web and social media, the movement swelled exponentially, and soon large-scale protests had broken out across the Middle East and North Africa, some of them comprising a million people or more. By 2011, more than a dozen countries were in the throes of revolution, including Libya, Egypt, Syria, Yemen and Iraq. The protests were coordinated and amplified on Facebook and Twitter. The use of social media platforms more than doubled in Arab countries during this time, with many of the users accessing the internet for the first time. Perhaps the most memorable congregation of people was in Tahrir Square in downtown Cairo. Beginning in late January 2011, several hundred thousand people gathered there to call for an end to the twenty-nine-year reign of President Hosni Mubarak. In February 2011, Mubarak resigned. He was succeeded by Mohammed Morsi, who in 2012 won Egypt's first democratic presidential election.

In the earliest days, at least, the Arab Spring looked to be a positive and overdue response to decades of tyrannical repression and economic stagnation. Four regional governments were overthrown, including Tunisia's. But while the internet proved to be an effective medium for organizing resistance, it was less effective at building democratic replacements. In Egypt, for example, Morsi served less than a year – he governed erratically and General Abdel Fattah al-Sisi took power in a coup d'état in 2013 and has held it ever since. In Libya, the death of Qadaffi led to a years-long civil war, followed by a delicate and highly tenuous truce. In Tunisia, where the protests took root, a true parliamentary democracy was established – only to succumb more recently to the return of strongman rule.

Overall, the Arab Spring was simultaneously one of the most inspiring and most frustrating political events of my lifetime. Technology is not a panacea; the mere existence of the web does not automatically result in the widespread establishment of human rights. And human rights themselves can mean different things; we in the West must not repeat colonial assumptions about the way things ought to be. But I would note that Europe's evolution from monarchy to democracy was hardly a straightforward affair either. It took centuries, with plenty of backsliding (and plenty of sectarian religious warfare) along the way. Human progress is not linear, but my sense is that it generally points in the correct direction. I remain optimistic that the web will continue to be a platform for democracy in this part of the world. But it depends how we use it – and what systems we build on it.

•

Many of the protesters in the Arab Spring were mobile-first – or often mobile-*only* – internet users. They weren't accessing Facebook or Twitter directly through a browser, but instead through dedicated, downloadable apps on their phones. This sometimes led people to conclude that 'the web' was being replaced by the app marketplace.

This wasn't quite accurate. The *browser* was being superseded by the mobile apps, certainly, but the back-end infrastructure that serves the content to these apps still runs on HTTP. Popular apps like X, Instagram, YouTube and TikTok all use the hypertext protocol to retrieve content. Spend the day refreshing your feeds and you'll make several hundred calls to HTTP servers in the process. For some services, the browser is even embedded inside the app. So, although the web may seem 'invisible' in these apps, it's still there – it's just running behind the scenes.

However, the popularity of dedicated mobile apps led to a shift in the user experience. On the PC, the browser is today by far the dominant application. Almost everything you do with a laptop or desktop computer – word processing, chatting, video streaming, even gaming – now runs inside a browser tab. On a mobile, however, the opposite is true; the browser is just one application among many, and typically not the most popular one.

This happens because providers guide users into the app ecosystem. Access Reddit or a similar site on the mobile web and you will continually be nagged about switching to the app. The app is more profitable for them, and allows them to track what you are doing – in short, it gives them a lot of control. It also makes it harder for the user to leave the service to look at something else. (Instagram won't even let you include links in a post, only the bio.)

I think the mobile experience would be greatly improved if you didn't have to download an app every time you wanted to engage with a new service. Interestingly, that's what Steve Jobs seemed to first showcase with the iPhone. If you watch his original presentation for the device, from 2007, he repeatedly demonstrates the unlimited functionality of the iPhone's Safari browser. The Apple app store wouldn't launch for more than a year.

Why did Jobs change direction? The answer was suggested to me at a clandestine rendezvous I had in the late 2000s with a Google engineer who shall remain unnamed. We met at a restaurant at Half Moon Bay, a popular beach town across the mountains from Silicon Valley. At a quiet table overlooking the ocean, the engineer explained to me that, from what he could see from deep inside Google, Apple was deliberately throttling the functionality of both the Safari and Chrome mobile browsers. Apple, you see, got a 30 per cent commission of the apps it sold; plus, it received continuing income from subscriptions and in-app purchases. It was far more profitable for Apple to direct the developers to build an app than to build a mobile website. Apple got a cut from the former, and nothing for the latter.

I considered what he was saying as I stared out into the Pacific. He was right, of course – Apple had found a way to profit from not just apps, but also magazine subscriptions, in-game transactions and anything else that was sold through its platform. Worse, when vendors tried to direct customers to pay using cheaper web-based alternatives, Apple kicked them out of the store! Some of the more powerful vendors objected; Spotify later filed an antitrust complaint against Apple, and Amazon to this day will not sell e-books through its Kindle mobile app. But few vendors had the clout to fight Apple, especially given the predominance of the iPhone.

I considered the implications of Apple's competitive position. The Apple app store was an example of exactly the kind of for-profit gatekeeping I'd always hoped the web could navigate around. But the iPhone was extraordinarily popular – I owned one myself. The toll booths the web had managed to avoid on the PC had found their way into the mobile realm.

•

Following the Arab Spring, Rosemary and I began to think about how best to direct the Web Foundation's resources. Rosemary had for many years been a fellow at the Berkman Klein Center for Internet and Society at Harvard, and so the intersection between the web and its impact on societies was front of mind. With help from Google's philanthropic arm, in 2014, we published the pilot edition of the Web Index, which ranked countries by their access, freedom, openness and relevant content. Denmark was judged to be the most open country, with a score of 87; Ethiopia, with its severe restrictions on content and its illiberal telecom monopoly, got a 34. Having these metrics available helped us to direct our policy work better, particularly in the Middle East and Africa, in the years to come.

An index of this type can become a roadmap for development. In assigning metrics, you're actually making value statements: Can you use the web to find jobs? Can kids use it to get an education? Can your government turn the internet off? We announced the publication of the index at the National Theatre in London, with the comedian and actor Stephen Fry as the MC. The brutalist theatre building is a nested series of concrete boxes on the Thames; Stephen joked that it has 'the best view of London, because it's the only place you don't have to see the National Theatre'.

Around this time, Rosemary also joined me as my partner to an Order of Merit luncheon with the Queen. Rosemary had in fact met the Queen once before, as a girl in Canada, when she'd been randomly selected to attend a function during a royal visit. Well, what do you know? And now, several decades later, Rosemary actually had lunch with the Queen at Buckingham Palace. (*She* wasn't forking any artichokes.) During our lunch, I got the impression that Her Majesty was sizing Rosemary up a little bit. Years later, in 2014, when I asked Rosemary to marry me, the Queen gave us permission to use the Chapel Royal, dating from Henry VIII's time, and the adjoining St James's Palace, as our venue.

The Chapel Royal had exactly 104 seats. We both come from large families and there was just enough room. The most important thing for us was to include our children – we were bringing two families together, and they all had a part. All of them were grown now, but over time they had become friends, and respectful of each other. Rosemary's children, Jamie, Lyssie and Indi, were very bright, and I now think of them, along with Alice and Ben, as our children, very much as a group of five.

Thrilled . . . Rosemary and me outside the Chapel Royal on our wedding day.

Rosemary also developed a great relationship with my mother; the two of them even went dress shopping together for the wedding. 'We found this quirky shop on King's Road, and I saw this really funky dress that I thought would look beautiful on her,' Rosemary recalls. 'She was eighty-eight years old by this time. I called her to tell her and she jumped in a taxi from Sheen on a Saturday afternoon. She put on the dress and just looked magical. We were so fortunate to have both of Tim's parents with us for the wedding.' The subdean of St James's officiated, wearing a large and splendid mitre. It was really a great wedding, a very special day.

My mother, Mary Lee, on the day of the wedding.

•

In 2011, I was elected to the board of the Ford Foundation. The foundation, originally endowed by family members of the Ford Motor Company, is today one of the largest private endowments in the world and gives away several hundred million dollars' worth of charitable grants every year. As a board member, I served a six-year term, during which I was able to learn more about the act of philanthropy. (The web had been given away for free, which could be considered a charitable act, but I had never given anyone a direct grant.)

One of the many things they did at Ford was to fund a group which would use geospatial mapping data to develop land rights for indigenous communities. Much of that work was done in Brazil, where communities that had existed for centuries were under threat from foreign developers who claimed to now 'own' the land. The Ford Foundation also worked with remote Afro-Brazilian communities which had been founded by slaves who had escaped into the jungle in the eighteenth and nineteenth centuries. Land ownership is always a contentious issue, but, by using newly available satellite and GPS data, these communities were able to have land rights granted for the first time.

Around the same time as I joined the Ford Foundation, I received an email from Danny Boyle, the director of *Trainspotting*, *Slumdog Millionaire* and *28 Days Later*. Danny was the artistic director for the Opening Ceremony of the 2012 London Olympics, and he was wondering if I might want to be involved. At first we thought it was spam. Why would the creative director of the Olympics want to meet us? I'm not a huge movie buff and I must admit that while his name seemed familiar to me, I couldn't immediately place who Boyle was. Only after googling him did I realize the email was genuine. We were curious about what he might have in mind for a computer scientist like me during a spectacle like the Opening Ceremony.

A few weeks later, we found ourselves in Boyle's office in the East End of London, just a short distance from where the Olympic Stadium was being built. After a warm welcome, Danny laid out his vision for the ceremony – a journey through British history, from pastoral times to today's digital age, including the Olympic rings being forged during the Industrial Revolution. He showed me a model of the set and explained that he wanted to take risks to deliver a portrait (and history) of Britain that would entertain, inspire and provoke. He was clear that while he wanted the evening to be celebratory, he would not shy away from difficult topics and that the whole thing would be infused with the quintessential British qualities of humour and self-deprecation. In his view, there were many connection points between the web and the Olympics – chiefly their shared ability to unite people across cultures, races and geographies. He asked if I would consider playing a role, to help celebrate the web as one of the great inventions that Britons had shared with the world.

I agreed, of course. In keeping with the historical theme of the ceremony, Danny wanted me to sit at a NeXT workstation, then type a short phrase that would be repeated in lights around the stadium. The phrase 'This Is For Everyone' just instantly felt right. It neatly encapsulated my vision and hopes for the web, and could be seen both as a celebration of the past while suggesting how much work lay ahead. And, of course, it embodied the spirit of the Opening Ceremony, and the Olympic movement as a whole. A day or two before the ceremony, a thought popped into my head. Since I was there representing the web, how could I ensure this message was shared online at the same time as on stage and on TV? Could I tweet it? A quick discussion with the production team followed – we all agreed that it was a good idea, but since the NeXTcube on stage was not going to be connected to the internet, we had to work out the logistics. In the end, we agreed on a low-tech solution – I would type the tweet out on my smartphone in advance, and leave it with a trusted production assistant, who would hit 'send' at the right moment.

In rehearsals for the Opening Ceremony, I started to find out who else would be participating, and what the day itself would entail. This really hit home for me at the first full rehearsal, when I walked through the dressing-room area in the depths of the Olympic Stadium. As 'talent' in the show, I had been assigned my own dressing room, and the names on the doors were like a roll call of 'who's who' in Britain in 2012, with stars from the worlds of music, literature, stage and screen all represented. James Bond was there, in the form of Daniel Craig; Mr Bean was there, too, in the form of Rowan Atkinson. Coldplay, Adele and the Rolling Stones were all scheduled to perform. I had a 'pinch me' moment when I stood in front of Paul McCartney's door (he performed the closing musical number for the ceremony). It was also wonderful for me that our children were able to attend rehearsals and the ceremony, as many of our family heroes and favourites were participating.

As the ceremony approached, Danny and I also discussed in more detail some of the difficult issues he would call attention to. One powerful example during the show was when the unintended consequences of the Industrial Revolution were explored. The Olympic rings were forged in this act of the ceremony, and while this was a moment of awe and progress, it also meant that Blake's 'green and pleasant' land was being torn up, with smokestacks, smog and noise (delivered via an intense drumming sequence) introduced into the world.

This theme resonated with me. I believe the web has been a massive positive force for humanity, but even in 2012, it was becoming clear that there were unintended negative consequences in what we were developing. As it had in the Industrial Revolution, the balance of power was swinging too far in favour of large corporations, not to mention governments, and away from individuals. The sense that the web was *extracting* value, in the form of personal data, from its users was something of growing concern for me. The sense that the web was *surveilling* users, rather than liberating them, was also a rising source of anxiety. As the ceremony approached, I was reminded again that we were engaged in a battle for the soul of the web.

The ceremony called for me to sit at a computer inside a house, then for the house to be winched up above me. There had been two full rehearsals, and a bunch of small ones. This would easily be the largest number of people I had ever been in front of, and there were so many ways that things could go wrong – I could trip over a shoelace, the house could fail to be winched up correctly, as had happened in an early rehearsal . . . the list went on! My nerves were set further on edge when, en route to the stadium, the driver of my vehicle briefly picked up his mobile phone and was immediately pulled over by a traffic officer for using his phone while driving. Luckily, we were let off with just a warning, but this was honestly the first time in my life I'd considered playing the 'Do you know who I am and where I'm going?' card.

Thankfully, from the time I got to the stadium, it was smooth sailing. I headed straight to my dressing room – spotting Emeli Sandé and Sir Kenneth Branagh from afar – and saw various other celebrities and their entourages in the green room. Time seemed to pass in a blur, and the next thing I knew, it was time for action. I headed through a production tunnel under the set, took my place at the desk and was seated inside a giant, fake house, waiting for my part to begin.

I was wearing a cream-coloured linen suit and an open-collar linen shirt with no tie. The costume department had done a good job – to be honest, it might be the best I've ever looked. Outside the house, dancers in colourful costumes streamed by, while the rapper Dizzee Rascal performed a few feet away. As the number ended, all the dancers streamed through the door of the house, right at me, each a totally different vision of delight and energy and colour as they exited the stage. Then the winches whirred and the house was hoisted up into the sky, leaving me seated alone at a desk, in front of a computer in the full glare of the stadium lights and accompanied by blaring house music. I typed out 'THIS IS FOR EVERYONE', hit return, and watched the phrase speed around the London Olympic Stadium in giant LED lights, to cheers from 80,000 people at the Opening Ceremony of the 2012 Summer Olympic Games. I knew that hundreds of millions more were watching live on TV from every corner of the world, and since the phrase was being tweeted at the same time, it would travel even further.

I had just enough time to watch my message light up around the stadium, stand, acknowledge the crowd in all directions, and desperately try to catch a glimpse of my family in the stands. Then it was all over. I had to switch from being at the heart of the action to being just a spectator – a dance I sometimes feel I am doing with the web. Back down the same tunnel I went, and then up to join my loved ones, where I was delighted to discover we had been seated next to Paul McCartney's family – my kids had heard the

McCartney family saying, 'Look, Tim's kids are excited for Tim to come on, just like we are to watch Grandpa.'

As we watched the rest of the show, I was buzzing with adrenalin, and both Rosemary's and my phones were buzzing incessantly with messages, calls, emails and tweets. Along with the other thousands of participants, I had been determined to keep my involvement a secret until now, and friends, family and acquaintances around the world were reaching out in surprise and to congratulate me. Rather than try to reply to them all, I chose instead to soak up the moment. I had made my statement about what *I* wanted the web to be, and everyone in the world had seen it.

CHAPTER 11

Open Data

Concerns remained. The industrial-scale harvest of user data was part of a growing and alarming trend on the web. The main technology was the cookie, the small data block that servers placed inside the browser's history. First-party cookies were an absolutely crucial part of the web; without them, you'd have to re-enter your password every time you visited a site. It was *third*-party cookies that were really the problem. These were essentially tracking devices that would follow your journey across multiple sites. Worse, as the 'third-party' name suggested, they were set by domains other than the one the user is visiting – so after your visit to nytimes.com, you might leave with cookies from doubleclick.net.

As the web grew, third-party cookies proliferated everywhere – especially at media sites that relied on advertising as a primary form of revenue. In 2010, a user visiting a newspaper website might navigate away, unwittingly carrying a dozen such spying devices or more! There were browser extensions that blocked third-party cookies, but most users didn't install them. The average user was only aware that the same advertisements for the shoes they'd googled a week before now seemed to be following them around everywhere.

Cookies were the most common tracking technology, but they weren't the only one. Many pages used 'invisible pixels' embedded in seemingly innocuous images to gather IP addresses. A similar technology known as a 'beacon' could track not only user data but also the movement of the mouse pointer throughout the page – even hovering over an image would be recorded. These mechanisms were often used in combination to create detailed profiles of user behaviour, preferences and interests, which were then used for targeted advertising.

Some social media sites – but not all – rely on targeted advertising or trading in user personal data as their business model. For these sites, as many have observed, users are not actually the customers. Users are the *product*, and the customer is an advertiser, who pays for access to that product. And the more data one can collect on a user, the more that product is worth. These user data profiles may be anonymous, but they are also extraordinarily detailed. Facebook has hundreds of different targeting options for advertisers. A 2016 *Washington Post* investigation detailed some of these options, which included:

- your location, age, gender, language, education level, and ethnic affinity
- your estimated income and net worth
- whether you have an anniversary coming up
- whether you recently got married
- whether you were pregnant

- whether you had a *friend* who recently got married or pregnant
- whether you were in a new relationship
- whether you were in a long-distance relationship
- whether you were no longer in a relationship
- whether you were a parent, and if so, what kind of parent you were (soccer mom, single dad, traditional)
- whether you were a conservative or a liberal
- whether you were a conservative *pretending* to be a liberal
- whether you owned a boat or a motorcycle
- whether you *wanted* to own a boat or a motorcycle
- where you went out to eat, how often you went out to eat, and how much you spent
- your bank
- your internet service provider
- your cable company
- your church
- your favourite TV show, your favourite video game, and your favourite beauty products
- your recent online purchases
- your sexual preferences
- your addictions
- how often you'd googled 'am I dying' in the past calendar year.

It had never been my intention to have any data collected about the user at all, and the original web tools I wrote contained no mechanisms for doing so. In 2009, I was invited to a roundtable discussion with UK parliamentarians in Westminster to offer my thoughts on the growing encroachment on privacy. 'We use the internet without a thought that a third party would know what we have just clicked on,' I said. 'Yet the URLs which people use reveal a huge amount about their lives, loves, hates and fears. This is extremely sensitive information. People use the web in a crisis, when wondering whether they have a sexually transmitted disease, or cancer, when wondering if they are homosexual and whether to talk about it.' I could see how tempting this kind of information could be to advertisers. 'There will be a huge commercial pressure to release this data,' I said. 'The principle should be that it is not to be collected in the first place.'

•

At the same time, I could see the power of having all this data available. In the late 1990s, I often went on ski trips with my son, Ben. We would take along our phones, as well as a GPS tracker and a digital camera. We would ski all over the mountain, then come back to the hotel room and download the GPS data and photos to my laptop. Of course, in 2024, your phone does all this for you – but back then it did not! Instead, I had written a Python script that converted the GPS track into a line on the map. Tracking the photos was more difficult, as in those days cameras did not include location data in the image. But they did record the time the photograph was taken, so, by looking at the GPS trace, we were able to determine our location at any given time, and match it to where the photo must have been taken. The end result was a wonderful illustrated travelogue of our skiing day.

So even in the 1990s, I was using the power of *linked* data – in this case, linking the data from my GPS with the data from my camera. Of course, I was only able to do this because I knew how to code. As another example, in the early 2000s I began using the accounting software Quicken to keep track of my expenses. Quicken, made by the company Intuit, had an arrangement with Citibank that allowed you to import your credit card data automatically. You could classify these bank transactions as expenses or not, then use Intuit's Turbotax software to automatically populate your tax forms. I used Quicken for several years until one year, when I had finished filing my taxes, a little box popped up saying, 'You don't seem to have a lot of insurance. Can I help you find some more?'

Boom. This was a watershed moment for me. My trusted helper, a laptop app which I had paid good money for, had suddenly revealed it was not on my side. It was on Intuit's side, and it wanted to sell me some insurance. It had violated my privacy – after all, it knew a lot about me, having just gone through my finances, and it was perfectly placed to sell me things.

I stopped using Quicken and moved over to using my own system, adding bits of my own Python code to the workflow and downloading the credit card statements straight to my laptop. After several years, though, my code downloading the bank statements stopped working. I called the helpline, and, a long time later, they called me back and asked me to describe what my credit card looked like. I said it had a big Quicken logo on it. 'Ah!' they said. 'That was an old deal with Intuit for early adopters, but now we have just signed a deal with Intuit, so that you can only use this data with Quicken.' So they had blocked my Python programs from accessing my own spending data online – but that's not how I want my world to work! I want to have a separate choice of bank, and my own accounting software. 'Do you know what it sounds like when someone cuts up a credit card into a bunch of pieces?' I asked them. 'This is what it sounds like!' Another little watershed moment.

Even in these early stages, I was adopting strong personal principles that I wanted to be able to use whatever software I wanted, to do whatever I wanted. Another principle came from Dan Connolly, my geeky colleague at MIT and W3C, who has as his mantra: 'The bane of my existence is doing things my computer could have done for me.' People who have that feeling, or, really, anyone who writes code, are always at a decision point: shall I do this, or shall I write some code to do this? Many wonderful bits of code have sprung from that urge – although admittedly a lot of deadlines were also missed while the person got distracted coding.

Dan often blogged about using code to make his life more smooth and efficient. Any code he made, he left a trail of breadcrumbs – links to the things he had found in his quest – for others to follow. Like a lot of great geeks, Dan tended to postpone administrative tasks. When the W3C staff got fed up with the fact that he was late sending them invoices, he asked exactly what they needed, then wrote some code which faxed it straight to their line. When it arrived it was perfect – except it was the size of a postage stamp!

As I took note of what existing digital systems could do and what they couldn't, I found myself reflecting further on how information could be structured and layered and put to new uses – in short, about what the 'semantic web' could be. At the bottom of this web, I wanted there to be a layer of data – *facts*. Above that I wanted a layer of logical relationships, classifying things such as where a photo was taken, whether a transaction was an expense or not, etc. And at the top, I wanted machine agents using this data to make useful connections and deliver new insights. I called this the semantic web 'layer cake'.

You could say that we were trying to make AI by building smarter and smarter machines by hand. We had all read Asimov's books, about robots as smart as people. We had read Arthur C. Clarke's *2001: A Space Odyssey*, in which the computer gets out of control. I had ploughed through all that stuff when I was a teenager. We could all imagine AI, but we didn't have it.

But this feeling – of 'I don't want to have to do anything my computer could do' – was one which would push us towards the computer doing *everything*. The machine agents could do your taxes by working out one formula on the worksheet at a time, but there was a push within the semantic web community to do more than that. Some people thought that if the computer had access to a large enough database of factual relationships, it could move from logical reasoning to common sense.

The AI pioneer Doug Lenat had created a project called Cyc (from enCyclopaedia) in which people encoded common-sense relationships between things like tables, chairs, fruit, human families, and so on. Cyc had knowledge at two levels: facts, like 'Apples are a type of fruit', or 'Alice is a human', and also simple logic, like, 'Anything which is a mammal has a mother.' From there, you could deduce that since Alice is a human, and a human is a mammal, then

Alice has a mother. Using these systems, Lenat hoped to make reasonable inferences about the broader world.

Unfortunately, the whole movement of trying to build AI using logic did not produce the system Lenat wanted, at all. This created pushback, especially from those funding the research, resulting in the lean times for the field known as the 'AI winter'. Scientists found they had to distance themselves from AI to get funding. In fact, at MIT, for a long time, the computer science group and the AI group were separate labs – they were only reunited as the CSAIL lab in 2003.

•

Despite the lull in development of the semantic web of the 1990s, by the late 2000s, I was beginning to see a way forward. I have always believed the web should be a two-way thing: a place to write, as well as to read. Richard McManus, a journalist and blogger, captured this spirit in a blog he named the 'read–write web'. It was also a big value with the W3C team: my original WorldWideWeb.app on the NeXT had been an editor, not simply a browser: an application, like a word processor, which allows you to create and modify material as well as just read it. At INRIA, the French national computer science lab, Vincent Quint and Irene Greif had built a hypertext browser-editor, where you could even edit diagrams. The W3C website was set up so that you could submit your own web page. And of course, Wikipedia is a great realization of this idea.

So for documents, we can set up a read–write web. But what about for data? The data in our systems comes from external sources like banks, GPS devices, cameras, and so on, and is then combined on our laptop or desktop. But if we want to correct it, or alter it, we can't: it's read-only data. Within the semantic web community, the idea of a 'bit of the web of data which you own' was becoming more and more desirable. So we built containers for our own data, and we called the containers Personal Online Data Stores, or 'PODS'.

To make a pod, all you needed was a web server, with a couple of additional features. Soon, we had produced a specification so anyone could make a pod server. We struggled to think of what to call this specification, ultimately settling on 'Social Linked Data', or Solid. The 'social' aspect was because we were excited about using these pods in apps where people could collaborate. Your pod was not a thing in isolation, but a tool where groups would meet and organize.

With Solid we had an ecosystem where your pod would sit like a blank canvas, or an empty Scrabble board, waiting for apps to write facts into it. Then as a developer you could write your app and let your users store whatever data they wanted in their pods. The pod servers never have to be updated to understand how to handle a new type of data – they don't care. In this way, pod servers are universal: they will work with any type of data, right out of the box. And the read–write web of data *must* be universal in that it must allow you to store any new knowledge about anything.

The difference between Solid and the semantic web was that Solid was less interested in inferring new relationships between things, and more in allowing the user to use apps that *write* data into a universal container. The GPS and photo data I had downloaded to my laptop while skiing was now being generated automatically by my smartphone. That was great, but the problem was that it didn't link easily to any of my other data. Instead, one group of data was siloed on my phone, while another was in my financial statements, while a third was with my health provider, and a fourth with the government.

As we conceived it, you should be able to collect all of this data and store it in the pods. Rather than writing your credit card transactions to some CSV file in a corporate database, your bank would write it directly to your pod. You would then be free to link this data to whichever tax calculator you were using, or whatever budget app you preferred. And everything could go in your pod! Your phone records, your Facebook relationships, your comment history on YouTube, etc. Rather than have all this stuff siloed off with different providers across the web, you'd be able to store your entire digital information trail in a single private repository.

In its earliest conception, we tended to present a user's stored data as a diagram, with circles and arrows. This was a little too abstract for most people, so some of my undergrad students at MIT built a *tabulator*, which was the first really functional Solid product. The main take-away from the tabulator project was a key breakthrough: even though semantic web data had been displayed in graphs, users wanted to browse the data just as they explored existing data on their computer, using files and folders. The tabulator would allow you to explore the information bit by bit. At first, it would just list the properties of one thing, like, say, a film's title ('Avatar'), its genre, and lists of directors, writers and producers. Click on one of the directors, and that line would open up to provide information about that director, such as name, age, movies directed, movies acted in, and so on. This was very intuitive and generated a lot of enthusiasm for the project. I began to see how something like Solid could be to the web of data what HTML had been to the web of documents: a universal default layer that anyone could access and use.

Solid, like HTML before it, would be an open protocol which app developers and businesses could access through custom software. The Solid layer would accomplish two things simultaneously. First, it would restore the privacy of the individual on the web, who would no longer have to worry what data was being generated about them, or who was looking at it. Second, it would unlock all manner of new functionality, by connecting data that had previously been stored in separate containers.

For example, there were obviously huge benefits to be realized by attaching your smartwatch data to your medical records. If you were looking for new shows to watch, or new media to consume, you might attach your browsing history to a content recommendation algorithm. If you wanted to get a mortgage, you could attach your spending habits to your application. Maybe a travel agent would even be willing to pay *you* for access to information about the countries you'd visited and the restaurants you preferred. Of course, all of this was optional; if you were a privacy-oriented person, you would never have to share any of this information, and even if you did, you'd automatically be able to see who had access to anything you did share.

In 2015, research on the Solid platform at MIT got a boost with a $1 million gift from MasterCard. Early on, we had a proof-of-concept product to show: refugees in Africa were given a Solid ID and could store all of their data in a pod. MasterCard saw, as I did, that it was potentially world-changing to put people in charge of their data on the web.

Now, one obvious roadblock to Solid is that Facebook and Google want to keep your data profile to themselves, so that they can sell it to advertisers. What MasterCard saw was that a world where people have all their data in pods is *overall* a more valuable system to be in – and people will gravitate towards systems where they see more value! Sure, Solid was something of a moonshot, but if it worked, it would unlock a huge amount of value. There would be markets for pods, markets for apps, and markets for people who did cool stuff with your data. MasterCard wanted in on the ground floor of that, and funding the work meant they could keep an eye on it. With the money from MasterCard, we continued to build the Solid layer, and soon had a prototype of the first Solid server up and running.

•

Even as some kinds of private data that shouldn't have been shared were shared, other forms of data that should have been shared weren't. Every year governments collect reams of data – in maps, bus timetables, government spending breakdowns, records of traffic accidents, environmental measures and economic statistics. This data belongs to citizens, as it is paid for with taxpayers' money. With the exception of sensitive personal data and matters of national security, most of that data deserves to be shared with the public and academic researchers. Governments are great at gathering information, but they are generally pretty terrible at analysing it. By properly formatting and sharing this data, a great amount of useful empirical social science might be conducted. Citizens would also have better access to the information their governments generated, which in a democracy was rightfully theirs.

Unfortunately, entrenched bureaucracies came up with a great number of excuses not to share this information. So in 2009 and 2010, I gave TED talks demonstrating the power of publishing open government data to the web. (In the first, I had the audience chant 'Raw Data Now!'.) Around this time, UK Prime Minister Gordon Brown invited Rosemary and me to Chequers, the prime minister's country residence, for lunch. Chequers has a very pleasant L-shaped garden at the back, and, as we ate, Brown asked us, 'What should the UK do to make the best use of the internet?'

'Put all the government data on the web for free,' was my immediate response.

'OK,' he said. 'Let's do that!'

Brown then asked if we might lead the initiative to do that, but there didn't seem to be enough time to do so between our work with the Web Foundation, W3C and Rosemary's board commitments. Rosemary suggested that Nigel Shadbolt, then a professor at the University of Southampton, take the lead. Meanwhile, I was able to secure from Brown a funding commitment for £45 million to develop a Semantic Web Institute that would make work in the field accessible and interpretable for researchers and the public.

Data.gov.uk launched in 2009, and both Nigel and I were eager to get to work. But then, in 2010, the Labour Party lost the election, and Brown was replaced by the Conservative Party's David Cameron, who immediately cancelled the project. Fortunately,

Nigel and I got to make our case to Cameron as well. He seemed a little uncomfortable talking to scientists; but he had the very bright minister David Willets assisting him. With Willets's help, in 2012, Nigel and I secured a grant of £38 million from the UK government to start the non-profit Open Data Institute (ODI). This was all coordinated on the civil service side by Gordon Brown's private secretary Jeremy Heywood, who, somewhat unusually, stayed on to become Cameron's head of the Cabinet Office. Jeremy was a wonderful, quiet doer of good. He sadly died of lung cancer in 2018, which was a great loss for the community and the country.

We consider a data set 'open' if it is free for anyone to use, analyse and reproduce. For this reason, W3C came up with grades for different data formats governments or others might use, rating them with one to five stars:

★ Make your stuff available on the web (whatever format) under an open licence.

★★ Make it available as structured data (e.g., Excel instead of an image scan of a table).

★★★ Make it available in a non-proprietary open format (e.g., CSV instead of Excel).

★★★★ Use URLs to denote things, so that people can point at your stuff.

★★★★★ Link your data to other data to provide context.

To encourage government ministries to publish their data in a structured, non-proprietary, linked format on the web using URLs, we had 'five-star data' coffee mugs made up and sent them around.

We didn't always get five-star data released, but it was at least a step in the right direction. The ODI secured office space in the Old Street area of London and hired a staff of around thirty. (We later moved to King's Cross.) We began working with policymakers to make all sorts of government data – tax info, census data, official government statistics – more open and accessible to citizens, journalists and analysts.

The work continues to this day. The ODI also helps governments build digital tools to interpret all this information. For example, we developed a visualization tool to help the UK government tackle the spiralling cost-of-living crisis, highlighting hotspots across the UK where the combined cost of fuel, food, housing and debt were making whole swathes of the country unaffordable for young people.

The ODI's work has great potential to keep citizens better informed about what their governments actually do. Legislation in both the US and UK guarantees that new government data is open and machine-readable by default. The ability to analyse these data sets allows for a much better understanding of what's actually happening in our countries.

For the UK and other advanced countries, the ODI's work often involves navigating large-scale government bureaucracies, some of which are centuries old, but in developing countries, this data is sometimes being digitized for the first time. Either way, transparency as a public good is an excellent foundation for democracy. In 2015, the ODI provided technical support for the first democratic election in Burkina Faso in over thirty years. A mobile web app, powered by an open government data portal we helped to design, delivered results from twenty-one districts within hours of the polls closing. The result was Burkina Faso's first-ever democratic transition.

•

The web community had a number of allies in our fight for open data. One of the most important was the activist and hacker Aaron Swartz. Aaron had co-founded Reddit, but he found corporate life uncongenial and had taken up the crusade for a liberated internet. He spent a lot of time at the MIT campus and I was one of his mentors. He attended semantic web meetings, and he'd often come for lunch in my office. He was lively, intelligent and connected, with wonderful enthusiasm.

Aaron's approach to opening government data was more direct than my own. In 2008, in collaboration with Carl Malamud, an advocate of the public domain, Aaron had downloaded much of the database of US federal court records known as PACER. Even though it was public information, you had to pay to access PACER, and Aaron thought this was unfair. Working at the public library, Aaron created a Python script that automatically pulled requests from the site, and, over a few weeks, he and others collected nearly 20 million pages of legal documents and placed them in a free online archive, called RECAP. The project drew attention from the US government and the FBI began an investigation into Aaron's activities, though no charges were ultimately filed.

I admired hacktivism of this sort, which achieved through direct action what lobbying did not. Aaron and Carl belonged to our growing group of protesters who felt this kind of information should be open and free – as, indeed, the law clearly dictated. The US Department of Justice had somewhat maliciously complied with the law by sticking PACER records behind an expensive paywall; a single case might cost several hundred dollars to access fully. The RECAP archive showed what happened when you forced governments to make things truly free.

Aaron was part of the local open-data community, which included myself, Carl and the renowned legal scholar Lawrence Lessig, with whom Aaron was very close. Aaron was also connected to members of the hacktivist collective Anonymous, who would conduct denial-of-service attacks against targets they thought were unethical. I did not agree with everything Anonymous did, but I appreciated that they were fighting a battle against big corporations and big government. Give me a choice, and I will take the side of Robin Hood over the Sheriff of Nottingham.

Following the PACER pull, Aaron, using an MIT guest login he'd been issued, began bulk-downloading articles from JSTOR, a paywalled digital library that holds the vast majority of articles and papers produced by academics worldwide. Aaron believed that academic research funded by public money should be freely available to everyone, but this was a more aggressive action than the PACER scrape and would end up having significant legal repercussions. MIT first organized a sting operation, catching Aaron in the act with a computer he'd stashed inside an unlocked cupboard and linked to the MIT network. Aaron reached a settlement with both the campus police and JSTOR – I thought MIT was overreaching and had argued that it should drop the case. But then federal prosecutors brought a number of disproportionate wire fraud (fraud committed via electronic communications) charges against Aaron, carrying the possibility of thirty-five years in prison. This for bulk-downloading academic papers! On 11 January 2013, after negotiations for a plea bargain collapsed, Aaron committed suicide in his Brooklyn apartment.

Aaron's death came as a massive shock. Shortly after, I posted a memorial poem for him on Twitter:

> *Aaron dead.*
> *World wanderers, we have lost a wise elder.*
> *Hackers for right, we are one down.*
> *Parents all, we have lost a child. Let us weep.*

I attended his funeral, where his family and the 'tech for good' community overlapped. His family was sitting shiva for him, and there I got to know the other people in his life. I hope his family realized what huge respect everybody had for him. Aaron is greatly missed.

CHAPTER 12

Machine Learning

In spring 2014, the web celebrated its twenty-fifth anniversary. The exact date I first submitted my proposal at CERN is lost to time, as the first document reads only 'March 1989'. But when you invent something, people tend to want a specific date to be celebrated, so I give them 12 March – which is actually my mother's birthday. As the years have gone by, I have enjoyed that little link more and more. Not everybody knows about it, but you do now.

A few weeks after the web's 'official' twenty-fifth birthday, Google hosted a congratulatory dinner for the Web Foundation at their colourful offices in London. Upon arrival, Rosemary and I were greeted by a series of enormous Union Jacks painted on the walls. We passed through the playfully styled recreational zones and lavish cafeteria before arriving at the staff breakfast and lunch hall, which had been cordoned off for our dinner. This was classic Google style, upbeat and fun, and I was grateful to them for hosting the affair. It turned out to be a momentous occasion in more ways than one, for it was at this twenty-fifth anniversary celebration that

I first met Demis Hassabis, one of the brilliant computer science pioneers behind the revolutionary AI start-up DeepMind and now a Nobel Prize winner.

Google had acquired DeepMind for $400 million a few months earlier. At the time this was an unprecedented sum for an AI acquisition, although in later years it would come to look like a bargain. At the anniversary, Google had scheduled me to talk first, with Demis giving the closing lecture. After a simple and delicious dinner, I gave a few remarks on the past, present and future of the web. Then I took my chair and listened as Demis began to outline the extraordinary breakthroughs the DeepMind team had made.

Demis was a kind and obviously extremely intelligent man. He spoke in a quiet and unhurried voice and was easy to understand, but what I appreciated above all was the way he could take complicated concepts and make them simple. I find that the smartest people are often the easiest to talk to.

Demis had been born in London, to a Singaporean mother and a Greek Cypriot father. He had excelled in school as a young man and had captained England's junior chess team. In his early twenties, Demis had for a time worked as a video-game developer, but this proved to be an insufficient challenge for his intellect. After receiving a PhD from University College London he went on to confront greater tasks. In 2010, along with Shane Legg and Mustafa Suleyman, Demis founded DeepMind, looking to build the world's first 'general' artificial intelligence, capable not just of playing video games or solving chess problems, but of tackling any task that a human could do.

At the dinner, Demis outlined the incredible progress his team had made in just four years. Rather than building logic atop a layer of facts, as the semantic web community had done, Demis had used a neural network, a type of software patterned after the biological brain. At DeepMind, his team had created a synthetic training environment to allow this neural network to learn 'by itself'. The first challenge was the simple video game *Pong*. The neural network was given the paddle controls and a view of the screen, and told only that getting points was good. It started out terrible, slowly got better, and after millions of games, played *Pong* like a master. Soon the same network learned a variety of other vintage arcade games, receiving no instructions save the overall objective of running up points on the screen.

I watched Demis's talk with great interest. Despite my interest in AI, I had never worked with a neural network. The AI winter of the eighties and nineties was long and brutal, and for many years it had been almost impossible to get funding for neural network research. I had my own research track, and I didn't really know the machine-learning people. Neural nets were interesting, but they had never been very powerful. It was Demis who showed me how such systems, which would learn *without* explicit logic layers, could in fact solve hard problems. As his talk progressed, Demis discussed the great number of recreational and scientific fields to which these neural networks might be applied. As the talk came to a close, he also sounded a note of caution – DeepMind was making such rapid progress with AI that there was a risk such systems might advance beyond human control.

The audience reaction was one of dumbfounded awe. When someone did manage to get a word out, that word was usually 'wow', although Rosemary recalls a few people also saying 'uh-oh'.

For many of us, Demis's talk was our introduction to the modern neural network paradigm. There seemed no question that AI, and machine learning in particular, had made a radical and unprecedented breakthrough. Later, we met with Demis for a private dinner in London. After he cautiously turned his cell phone off, he expressed to us his concerns that the world was fragmenting into competing AI spheres, one centred on the West and the other in China. The research communities were feeding the AIs information about individual people. Given government regulations, Google had to jump through a lot of hoops to feed this information to their AI, but China didn't have that hurdle. He was concerned this might allow China to take the lead.

I was worried at that time that AI might become a strategic military issue. Most of the time web technology lent itself to building a shared environment for people to collaborate. AI, however, had a lot of weapons-grade applications, creating the potential for an arms race. In China, the Great Firewall was already in place, and a lot of sites were blocked. (Nor was China the only country doing this.) In my conversation with Demis I saw the potential for more offensive capabilities. AI could produce weapons, tools to probe cybersecurity, and the ability to breach another country's defences. All of these concerns have proven well warranted.

But if there were concerns, there was also optimism. I was (and am!) tremendously excited about the potential for AI. Over the years, Demis and his team have taught their neural networks to do increasingly powerful feats, graduating from simple arcade favourites to complex games of abstract strategy. In 2016, the AlphaGo neural network became the first computer to defeat a human world champion at Go. In 2017, its successor AlphaZero emerged as the most powerful chess computer in existence. Concurrently, Demis was teaching his machine to tackle the tricky 'protein-folding' problem. Protein folding is the process by which chains of amino acids form into three-dimensional shapes. Biochemists looking to speed the development of new medical treatments wanted to model this molecular origami, but computers had struggled with the problem. Demis and his team, building on their insights into chess and Go, produced AlphaFold, a neural net which predicted protein shapes with unprecedented accuracy. In the past, it might take a graduate researcher their entire PhD to figure out a single protein. AlphaFold could do it in minutes. For this, Demis was awarded the 2024 Nobel Prize in Chemistry, and, a little more than ten years after he delivered his Web Foundation anniversary lecture, we were once again invited to a celebratory dinner hosted by Google. Now it was *our* turn to congratulate *him*!

Demis's success led me to reflect on my own research. What we might call the 'facts and logic' approach to AI has produced some useful tools. Apple's Siri came from this approach, as did IBM's Watson, which had been programmed with facts about medical data and *Jeopardy!* questions. There was also Stephen Wolfram's 'Wolfram Alpha', which is a great resource for maths homework, giving both logical and also beautiful graphical results when asked a general question about advanced maths. These systems had all been built deliberately, with explicit knowledge bases and rules, and with more or less complete lists of things the machine should be able to answer. Some call this arc of development 'Good Old-Fashioned AI', or GOFAI for short.

Machine learning works differently. It has three legs: software, hardware and data. The software is the neural net, and the various techniques used to train it. The hardware is a special kind of microchip known as a Graphical Processing Unit (GPU), which can process calculations at incredible speed. (Nvidia, who makes such GPUs, is, at the time of writing, one of the most valuable companies in the world.) It is the third leg, the data, that is in many cases the missing piece of the puzzle. For this reason, I see the semantic web – the web of data – as being complementary to the work Demis is doing. Having a world of carefully organized data is very helpful to AI, whatever approach you are using. To solve the protein-folding problem, for example, Demis had trained AlphaFold on vast libraries of known amino-acid sequences and their related protein structures, which acted as the practice questions and answer keys for the biochemical curriculum.

There remained a vast ocean of data out there from which this new form of AI might learn. Web data, in particular, represented an open frontier and, a few years after Demis's 2014 talk at Google, the OpenAI team began to train powerful new language models using almost the entire web as an input. (More on that in a bit.) There was also your *personal* data, which might be employed in all sorts of powerful new AI tools, but only if you trusted the systems in charge to use it.

•

Travelling the world with my smartphone in hand, I was by now convinced that transforming the storage and use of personal data was the key to saving the web and improving our digital lives, and that Solid offered an excellent way to do all this. I had an airline app for each airline I used, but they didn't talk to each other. I couldn't look up a British Airways booking in the American Airlines app, and I had to tell them separately about my preferences and my passports – plus enter this information again, any time I took a new airline. Then there was the Customs and Borders Protection app I use when I land in the US, which of course didn't talk to the airlines. I had an Apple wallet which took credit cards, boarding passes and train tickets, and even a coupon for a meal at Club Passim in Cambridge – but the wallet didn't take my railcards, which I had to show on a different ticket app. And while Apple Health did a good job of importing fitness data, it didn't have my vaccination cards. Really, I thought to myself, what's wrong with this picture? How have we ended up in this situation? An alien coming to Earth would think it was very strange that I had to tell my phone the same things again and again – things it had just told me!

One reason the apps on my phone didn't talk to each other is that we basically don't trust our apps. The fear, which is a valid one, is that some app on my phone or on the web is going to attack me and harm me using my data. So we can't just let them all talk to each other. Another reason the apps didn't talk to each other is that they *couldn't*, because they didn't speak the same language. If two people each write an airline app, they will store pretty much the same information about flights and seats, but they will write it in a way in which they can't read one another's data.

If you've got this far in this book, you'll know that this is where *standards* come in. You'll know that standards are where developers from different companies and different projects come together and work really hard, in good will and good faith, to make a common language. Under these circumstances, each of their apps can communicate, and you can use different apps at different times – or at the same time! You can use any web browser, because they all speak the same hypertext language: HTML. You can pick any podcast reader as they all speak the same language: RSS. You can pick any calendar app because they all speak the same calendar language: CalDav.

With calendars, in fact, we are already more or less in the world that I want to get us to. On my calendar app, I can mix in data from quite different parts of my life: my personal, very private calendar, calendars I share with my family, ones I share with my friends, public calendars from my favourite band, my favourite sports team. And when you ask me whether we can have dinner tonight, the app pulls all that in and displays it so I can make an informed decision. I want that power for everything else, too. It's obvious what we have to do. Make standards like CalDav, but for everything.

That's what the Solid protocol does. Solid also deals with login and permissions, and functions as a kind of universal language: you can store anything on that data pod, including tickets, contracts, vaccinations, credit card transactions, exercise data, flights, diagnoses, genomes. Now, as the amount of data you'll generate in this way is overwhelming, in the near future you'll want AI to manage this data and make connections for you. Of all the apps you use, obviously a powerful AI is perhaps the most important to be able to trust.

But when I was first considering the links between Solid and AI, looking at the preponderance of (then relatively simple) voice agents like Alexa, Siri, and others, it seemed they all worked for big tech. That is, they were constantly acting as agents for airlines, hotels, and so on. I wanted an AI that worked for *me*. In 2017, I wrote for a talk – which became a note on my Design Issues hypertext page – an imagined conversation with such an AI, which I called Charlie:

> Bob was fed up with the AIs around him who all seemed to work for other people, and so he got Charlie. Charlie is an AI, and Charlie works for Bob. Because Charlie works for Bob, Bob gives Charlie access to much more data than he would another AI.
>
> – Charlie, who do you work for?
>
> – I work for you, Bob.
>
> Good morning, Bob. Good to see you on the exercise bike. Your fitness goals are on track. In fact because the meeting was moved do you want to stretch this to the full hour? We can do some climbs and have time to unwind.
>
> – OK, sure.
>
> – OK, so let's start at 100 cadence, to warm up, and we can go over a few things. Overnight a bunch of offers came in for your art, but as far as I could tell none of them really make sense to you after you've paid the fees. I just invested a little in one new start-up, mainly because it will give you something else in common with your mother-in-law. Speaking of relatives, you have quite a bunch of vegans coming on Saturday. I took the liberty of making up a recipe for the thing you really liked at the Indian cafe the other day.
>
> – You made up a recipe?

– Well, the chef hadn't published that one, but he has published a dozen books, so I read those as a training set and then extrapolated how he would cook the menu you liked. Then I compared it with the Linked Open Recipe data, and adjusted it a bit for the way you like things. So I propose to get the food from Whole Foods, Waitrose, and the Farm – we can get the best stuff and save 12% on the bill. OK?

– OK, Charlie, let's go for it.

– OK, the recipe is in your calendar. I'll leave you to your workout now. When you are done, there are two things: a new briefing for your meeting today, and the upcoming family birthday presents. I've found a bunch of things but I'm not sure they are right – I want you to look at them. OK?

– OK, Charlie. Who do you work for?

– Legally, ethically and algorithmically, I work for you, Bob.

Of course, in 2017 the power of AI actually to have that conversation was, I imagined, many years away – not realizing large language models like ChatGPT would soon be a thing. In fact this is the *near* future; such systems are right around the corner. But if we are to prevent these systems from exploiting us, it is critical that we get the data layer right. We need a layer where we control our own data, and we can share anything with anyone, or any agent – or no one.

•

In 2016, nearly forty years to the day after I graduated, I returned to Oxford University – this time as a research professor at Christ Church College. Returning to Oxford was a great personal honour for me, and I was really happy to come back. I revisited all my old haunts – the coffee shop (still there, but under new management), the wide greens and narrow passageways, and of course, the rivers.

Oxford had been a great part of my education and early life, and those three years as an undergraduate had been very special. It was a privilege to re-enter academic life. One of the great things about Christ Church is its interdisciplinary nature. When I go to lunch there, I never know who I might be sitting next to: a molecular biologist, a scholar of Old English, or a geographer with theories about the formation of modern towns. At one meal, I was seated next to Roger Davies, a professor of astrophysics who told me excitedly, and very articulately, about all the new exoplanets we were discovering in orbit around distant stars, just from studying the flicker of starlight. Over time, Roger Davies and I became good friends.

My new office at Christ Church had a second-storey view across a tranquil meadow to the River Thames. Walk around to the opposite side of the building and you could look down upon the dean's walled, private garden, which housed the large and unusual-looking 'Jabberwocky Tree'. Charles Dodgson, a professor of mathematics at Christ Church, had looked at these same trees, and they inspired his poem, which he wrote under the pen name Lewis Carroll. The dean's daughter, Alice Liddell, often played in the garden, entering through a small, secret door in the wall. She became Alice in Wonderland.

At Oxford's Computer Science Lab, I was able to invite students to participate in building out the Solid protocol. A key challenge was how to get insights about a group of people using Solid pods without disturbing the privacy of the individuals. In some ways, nothing had changed with the younger generation. They were all online of course, they all had phones, but they were at that special time in their lives with limitless potential. Now, I can tell if someone groks the potential of a technical proposal by the reaction in their eyes. With Solid, often people's eyes go flat, but with maybe one person in thirty, their eyes begin to twinkle. They *get* it. With the undergrads I worked with, that proportion of twinkling was much higher. They were sponges for new ideas.

I began to split my time between Cambridge, Massachusetts, and Oxford, England, moving into a former rectory. English countryside living suited me well, and Rosemary adored it too. As ever, I sought to acquaint myself with the place by running, and soon I had a regular route that took me through tilled pastures, rough brambles and half a dozen quaint but well-preserved villages. The end of my route went past a centuries-old village pub, with moss accumulating on a sloping rough-hewn roof. I often returned with my shoes soaked with dew and my shins torn from brambles.

I was at the old rectory late one evening in 2016 when I received a phone call informing me that I had won the Alan Turing Award. Presented annually in Silicon Valley by the Association for Computing Machinery, the Turing Award is sometimes also called 'the Nobel Prize of Computing'. It was a tremendous honour, and Rosemary and I went for a walk through the village and looked at the stars. Of course, first thing the following morning I phoned my parents. No one understood the importance of this award better than them – they had known Alan Turing personally. Both of them were delighted for me, my ninety-five-year-old dad especially.

Now, in my house growing up, when you accomplished

something special, you got a clipping placed on the kitchen refrigerator with a magnet. My parents took an egalitarian approach to recognition, and while they were alive, the four of us children shared fridge space equally. The Turing Prize was enough to get my photo to the front of the fridge, but it gave me no special privileges.

The award was presented a few months later at a banquet dinner in San Francisco. As you might imagine, it was a very geeky affair – not quite the Oscars, although everybody in computer science was there. Many past Turing Award-winners attended, including Vint Cerf, who was turning seventy-four that day. We opened the festivities by singing 'Happy Birthday' to him. After I was presented with the award, I delivered my acceptance speech, which – well, which ran a little long. 'When Tim's brain is running it goes so fast that he doesn't complete his sentences,' Cerf recalled. 'His ability to generate another concept in the next 30 milliseconds is incredible, but once he gets wrapped in something, he's totally unconscious of what's going on around him.' After I'd been speaking for about twenty minutes, Vint whispered in my ear. 'You're holding up dinner,' he said. I quickly wrapped up my remarks.

The trophy was a handsome silver bowl, which I added to my growing collection of prizes. Some of them seemed to have been invented solely for the purpose of getting the recipient to show up at an awards dinner. Sitting through awards dinners can get pretty tedious for Rosemary after a while. As my mother said, 'There should be a prize for the person who has to sit through the prize ceremony.'

•

Being at Oxford kept me in contact with the UK's burgeoning artificial intelligence scene. In fact, the next person to join the computer science department at Christ Church was Yarin Gal, one of the top machine-learning people. By the mid-2010s, machine-learning techniques were spurring a behind-the-scenes revolution in the way internet content was delivered. Machine learning was being used to connect listeners to new musicians and shoppers to new products, using a technique called 'collaborative filtering'. I first came across collaborative filtering while using the music-recommendation service Last.fm, which had been spun off from Southampton University in the UK in 2002. Last.fm 'scrobbles' my music library to keep track of my listening habits. This information is then compared against a database of users, where my habits are matched with clusters of similar listeners. So if I listen to a lot of Paul Simon (and I do) Last.fm will look at the habits of Paul Simon fans to see what other artists they liked, and use this information to make predictions about what else I might like to listen to – perhaps some band I'd never heard of until Last.fm recommended them to me. The more listeners who use this service, the smarter this prediction service becomes.

The ability of collaborative filtering to find you something you like is great. Spotify, following the lead of Last.fm, uses it to automatically generate playlists and radio stations customized to user preferences. Amazon and other online retailers incorporate the collaborative filtering technique into their product recommendation services, greatly improving the online shopping experience. Instagram, Pinterest and other social media sites use collaborative filtering to provide personalized recommendations.

What I was less excited about was the ability of collaborative filtering to find what agitates you. I saw this happen on platforms like Twitter and Facebook, which were optimizing for *engagement*, rather than enjoyment. It turns out that if you feed people provocative content, you can keep them on your platform for longer. So the AIs started showing nastier and nastier stuff – 'rage bait', basically – which made users angry and argumentative, but kept them logging on.

Over time, this collaborative filtering contributed to deep political polarization across many societies. Legacy media platforms on the web began to lose attention share to the social media companies which thrived on toxic discourse. Soon, political operatives realized they could use this new form of engagement to their advantage. The first indication that something was seriously wrong was the UK's decision to exit the European Union. The pro-Brexit crowd sold their position to the UK public on false terms, promising that exiting the EU would reduce immigration, improve the UK's economic standing and help balance the UK's budget. The deceit of the Brexit campaign was not itself unusual. What made the Brexit campaign different was *how* it misled the public.

The classic political misrepresentation is repeated ad nauseam on stump speeches and TV sound bites and even buses – in other words, it is *broadcast*. Everyone can see what is happening, and the countervailing truth can be broadcast to combat it. But in addition to this, the Brexit campaigners did something very different; rather than just broadcasting misinformation on TV, they *narrowcast* it via social media to the segment of the public that was most susceptible to their message.

The Brexit campaigners were assisted by the British political consulting firm Cambridge Analytica, who were experts in using Facebook data to target voters. Cambridge Analytica initially denied being involved with the Brexit campaign; it was only after the leak of a ten-page document entitled 'Big Data Solutions for the EU Referendum' that their participation was definitively revealed. Cambridge Analytica's strategy involved identifying and 'micro-targeting' persuadable voters based on 'advanced analytics to assign values on particular traits to the entire voting population of the area in question'.

Demographic profiling of the electorate was also nothing new, but the vast amount of data harvested on users resulted in far more detailed and accurate profiles than ever before. This could then be combined with 'A/B Testing', where user response rates to variations on the same political messaging were tracked to find the most effective headlines. The combination of microtargeting with misinformation resulted in effective, personalized political propaganda, delivered in secret to the segment of the electorate most predisposed to believe it. I'd add that another important factor in the Brexit debacle was the dysfunctionality of Article 50 – the clause in the Treaty on European Union that allows a member nation to initiate a withdrawal from the EU. Perhaps if there had been a Brexit vote *after* the negotiations as well as before then it would have been clearer what we were voting for. Instead we had to vote on what the politicians promised – and, as a result of narrowcasting and microtargeting, those promises could be different for different individuals.

The Brexit outcome had many causes, but a new political messaging system was surely one of them. However, I don't think

I really understood how bad that system was becoming until the US election of Donald Trump later on in 2016. Here, too, Cambridge Analytica played a role, working out of a second-floor office in London to influence the American electorate. 'Today in the United States we have somewhere close to four or five thousand data points on every individual,' Cambridge Analytica CEO Alexander Nix told SkyNews a few weeks before the election. 'So we model the personality of every adult across the United States, some 230 million people.'

Looking back, perhaps the influence of Cambridge Analytica was overstated in the immediate aftermath of the election. It is easy, though, to see why it caused such controversy. Traditional broadcast media had fairly strict regulations regarding political advertising; social media had few, if any. Furthermore, as the messages would only be seen by the small sliver of the population judged most receptive to them, it was very hard to track what was really going on. You didn't have to be a political scientist to understand the damage that provocative, targeted, secret and highly individualized propaganda could do to democratic discourse. I compare it, today, to an undercover megaphone.

I was in America for election night in 2016. A friend had organized a watch party at his house outside Boston. Following the polls, I had expected Hilary Clinton to win – as such, I'd brought along a bottle of champagne for the occasion. The watch party was festive to begin with, and many were excited that the so-called 'glass ceiling' was about to be shattered. As the results began to stream in, the room grew quiet, and the looks on my friends' faces turned from giddy anticipation to quiet disbelief. As the night wore on, it became clear that Trump, while losing the popular vote by quite a significant margin, was going to succeed in winning a majority in the electoral college and secure the presidency. The mood at the party grew stony, and people began to leave. I slunk off to the side and hid my bottle of champagne behind a curtain. Perhaps it's still there.

CHAPTER 13

Design Issues

I had known for a while that something was wrong with the web. What was intended to be a tool for creativity and collaboration had become divisive, polarizing and toxic. I had often talked publicly of the two Cs of the web: creativity and collaboration. After 2016, I began to talk of a third C: *compassion*. I was greatly concerned that the human element of the web was beginning to disappear. In its place we had large, faceless systems which spied on and manipulated the user.

Ever since my time at CERN, I have been recording my thoughts about the challenges facing the web on a public hypertext page entitled 'Design Issues'. I have been continually contributing to this web page for thirty-five years – I consider it the world's first blog (you can find it here: https://www.w3.org/DesignIssues/). The order of the posts is a bit scattered, and the language is fairly technical, but this approach is driven by what I'm trying to say. 'Having started this set of notes in 1990 in the (for me) novel medium of hypertext, it has been difficult to tear free of it: my attempts to lend hierarchical or serial order have been doomed to failure,' I wrote in a 1998 post. 'Neither have I found it easy to restrict myself to separated technical or philosophical arguments – and somehow this is, I feel, also important, the sharpening happening, after all, where the knife meets the stone.'

Following the political upheaval of 2016, these comments seemed more relevant than ever. The web was lacking in compassion, but this was not a human failing. It was a *design issue*. There was something wrong with the technical side of the web that was encouraging this toxicity. I identified two symptoms, stemming from the same disease.

Currently, the most egregious symptom is polarization. Social media, as currently built, leads users to take extreme political positions and demonize the opposing side. This makes constructive engagement difficult, allows outlandish conspiracy theories to flourish, and promotes demagoguery over deliberation. Soon, civilized discussion about important issues becomes impossible. Polarization, I fear, might have dire outcomes for humanity, with consequences on a global scale.

The second symptom is more individualized. Many social media users report suffering mental health issues after prolonged usage. The catalogue of ills related to social media is alarming: anxiety, depression, jealousy, inadequacy, feelings of isolation, body image issues – even suicidal thoughts. This mental health epidemic is especially acute among young people.

What is the common design issue that leads to these unfortunate symptoms? Web scientists analysing the information sphere have concluded that there is a direct link between social media and polarization. Social media companies are using machine-learning techniques to make users addicted to their platforms. These systems are *designed* to be addictive, feeding people more and more extreme content, making them alternately angry and sad. Unfortunately, there is a quirk in human psychology that draws us to these negative emotions – a quirk that the social media giants have learned to exploit.

Now, the desire to draw and retain users is as old as the web, and it can lead to wonderful things. During the early days of the web, users would put 'hit counters' on their blogs, which showed how many times the page had been visited; they would also link to other related blogs in a collaborative attempt to drive traffic. Excited by the thrill of someone reading your stuff, you aimed to make your work interesting and valuable, and you tried to link to other good people. From this, the interconnected network known as the 'blogosphere' developed, broadening the universe of discourse and allowing a great many new voices to be heard. The blogosphere emerged organically, and I didn't anticipate its development. It remains one of my favourite parts of the web.

You can draw a direct line from the hit counters to the 'like and subscribe' culture of the modern web. When social media began to take off, in the late 2000s, its growth at first resembled that of the blogosphere. Early users of Facebook, Twitter and Instagram often remarked on how *fun* these platforms were. But the algorithms that organized and presented content on these services prioritized engagement over enjoyment, collaboration or mental health, and soon the fun was gone, replaced by the signature act of 'doom-scrolling' through endless streams of invective, propaganda, negative news, conspiracy theory and envy-inducing lifestyle bait that preyed on the user's fear of missing out. In this way, the social media giants began to harvest users' attention on an industrial scale, vacuuming up a large portion of all web traffic.

One of the first to sound the alarm about these harmful but effective algorithmic practices was the software engineer Tristan Harris. Working at Google in the early 2010s, Tristan argued to his colleagues that software designers had a responsibility to ensure humanity did not spend much of its collective waking consciousness in a state of smartphone-induced anxiety. Tristan left Google to found the Center for Humane Technology and produced a 2020 Netflix movie about this issue called *The Social Dilemma*, which I highly recommend. The film presents the fictionalized narrative of two American teenagers: Ben, who is radicalized by provocative online political content, and Ilsa, who develops depression through her constant exposure to unrealistic beauty standards online.

There are a lot of Bens and Ilsas out there. In June 2024, Vivek Murthy, the surgeon general of the United States, called for a warning label for social media, similar to the warning labels placed on cigarettes. In an editorial in *The New York Times*, Vivek cited several grim statistics. 'Adolescents who spend more than three hours a day on social media face double the risk of anxiety and depression symptoms, and the average daily use in this age group, as of the summer of 2023, was 4.8 hours,' Vivek wrote. 'Additionally, nearly half of adolescents say social media makes them feel worse about their bodies.'

Unfortunately, Tristan's movie, like many criticisms of social media, only points out the problem. It does not mention what to do about it. It does not dive in at the point the algorithm is being trained and ask: 'Suppose instead of optimizing for time spent on the website, we instead tweak it to optimize for constructive engagement. Let's imagine what that small change would do.'

I agree with recent comments made by Yuval Harari, author of the books *Sapiens* and *Nexus*: 'If a social media algorithm recommends to people a hate-filled conspiracy theory, this is the fault not of the person who produced the conspiracy theory. It's the fault of the people who designed and let loose the algorithm.' Yuval has called for such algorithms to be regulated by the government. While I generally oppose the regulation of the web, in this instance I agree. My feeling is that regulation should be minimal, and only used when absolutely necessary. But, as I write this book, it is clear that, among all the wonderful things on the internet – and even all the not-so-wonderful things – there is a particular phenomenon which is causing harm.

I want to be clear that I do *not* think we should get rid of social media in general. Social media is a fantastic innovation with tremendous potential to bring people together. And given the vast amount of information uploaded to social media every day, we need algorithmic agents to organize the media that we see. All we need is to regulate the *addictive* algorithms, the ones that guide people into rabbit holes. It's the technical design decisions of the social network sites that drive this kind of polarization that we're targeting – it's the algorithm that's been trained to produce this outcome. We need to change that, one way or another.

•

I've been thinking about this problem for some time. In June 2024 I posted a diagram to my Design Issues page, which I called 'A Map of Everything on the Internet'. At the top were the data layers and protocols. These fed into the systems we all use, like email, the web, calendars, etc. On the right side of the map were all the good, open-source transparent systems, coloured in green. On the left side of the map were the harmful systems, coloured in red, and flowing into a menacing diamond entitled 'Feed Manipulation for Engagement'. A black-and-white version is pictured opposite.

People use the internet in a lot of ways which are actually helpful, benign, and even lead to good mental health, mutual support, collaboration on tough problems and unbelievable creativity. Apart from the web, they send emails, they read blogs and listen to podcasts. On the web, they dive into – and occasionally edit – Wikipedia. They use maps and weather sites, they share photos with friends and family, they buy stuff, they sell stuff. They do a lot of things which are useful, at school and at home, and not harmful. So there is a strong argument for providing a phone which has the good bits, but not the bad.

I drafted this map in response to a wave of people asking me if their kids should be allowed to have a smartphone. My answer to this is basically, yes. Kids should learn to collaborate, and they should learn to collaborate online with their friends. What we have to block is not smartphones, nor even social media, but the harmful algorithms often used by social media – the psychological equivalent to giving kids access to a slot machine!

Politicians in Australia recently proposed a ban on social media for kids under sixteen. While it's great that Australia is realizing there's

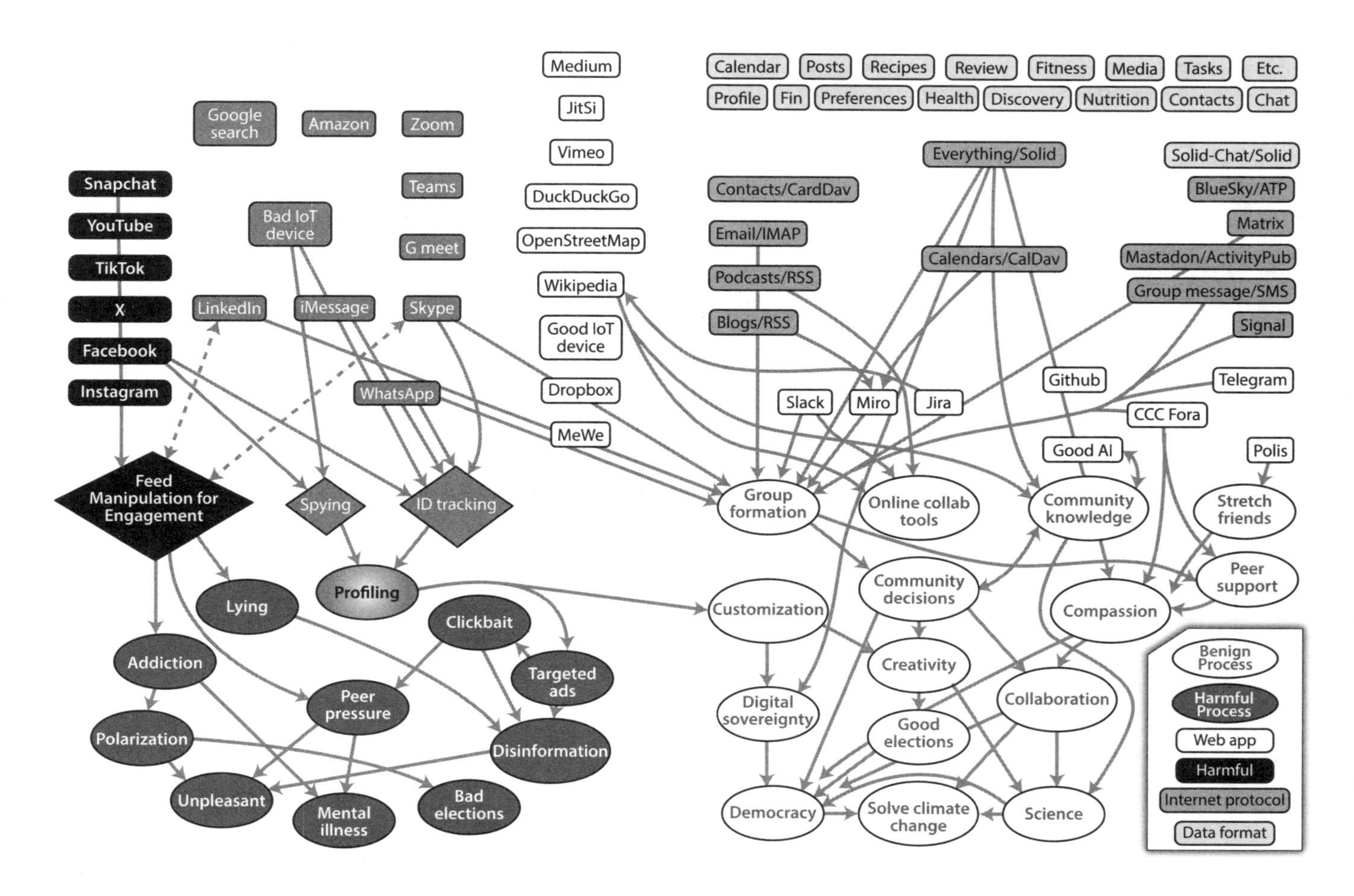

A diagram I created to visualize everything on the internet, showing that the bad causing harm to our society is only a small fraction of what is out there – so much of it is good.

an issue with children doom-scrolling in their classrooms, a ban is a bit draconian and deprives kids of a useful tool. What the Australian government should do is partner with phone manufacturers to create devices that block that very small (but ultra-invasive) part of the web that causes mental health problems and polarization.

These can then be replaced with better platforms. One such platform is the microblogging protocol Mastodon. It's my preferred replacement for Twitter, as it's completely open-source and runs across a distributed network instead of a centralized server. Mastodon's feed is structured on the timeline of events and the people you follow, without other more corrosive elements. As it is not funded by advertisements, it is not incentivized to hook teenagers into addiction. Without this incentive, the service becomes much less corrosive to mental health.

We can even go a step further. What if we trained algorithms for *constructive* engagement? What if we trained algorithms to maximize the joy people got out of discussing things with others, rather than feeding off hatred? Such algorithms already exist: for example, Spotify and Apple Music's algorithms do a pretty good job of selecting music that listeners will enjoy. Another site that seems to do a better job of promoting positive content is Pinterest. With such thoughts in mind, in spring 2024, the American businessman Frank McCourt organized a consortium. His aim was to buy the US assets of the video-sharing platform TikTok.

Frank is a wealthy Massachusetts real-estate developer who has decided the best thing to do with his time and fortune is to fix the internet. He founded Project Liberty, an NGO that 'builds solutions that help people take back control of their digital lives'. Frank, like me, is concerned about the damage that social networks are doing, which he sees as a result of their ownership of what he terms the 'social graph' – the unique set of connections with other users that we build through our interactions online. Frank, like me, is also concerned that the lack of interoperability between social media systems means that users can't import their social graphs from one platform to the next.

For this reason, Frank wants to develop alternatives to the eye-ball-maximizing approach that TikTok has used to capture the attention of the world's teenagers. He wants to turn TikTok into something calmer, something more like Pinterest. His thinking is similar to my own: let's not ban this service. Let's instead develop algorithms which maximize the TikTok users' quality-of-life, not their time-on-device. TikTok actually has the potential to do a lot of good, and, with this in mind, in May 2024 I endorsed Frank's bid.

•

Of course, adults are susceptible to the toxic effects of algorithmic feed manipulation, just as teenagers are. This negative engagement cycle has poisoned much online discourse, and, in the US, this has led to a broken political environment that just a decade ago would have seemed impossible. By promoting the most 'engaging' content (which, remember, is actually the most provocative), social media's feed manipulation algorithms can even turn outlandish lying into a winning strategy.

This is a global problem. In the 2022 presidential election in the Philippines, Bongbong Marcos, son of deposed president

Ferdinand and Imelda, eschewed traditional journalism to build his own social media ecosystem. Marcos used Facebook and TikTok to whitewash the human rights abuses of his parents' authoritarian regime, as well as to blame both the CIA and the Vatican for their downfall. 'The liberal weapons of historical accuracy and fact-checking are simply no match for Marcos' creative folklore, turbocharged by social media fan culture and relatable influencers,' wrote Jonathan Corpus Ong, an academic who studied the Marcos disinformation network.

We need to make social media that works for democracy. I'm encouraged by work done at the MIT Center for Constructive Communication, who, in conjunction with the non-profit Cortico, recently launched Fora, an open-source conversation platform. The Cortico team hosts small-group in-person discussions on local political issues. The Fora app records everything, but the recording is set as private by default. If the participants decide something someone said is important, they can promote it publicly using the Fora tool. The app produces two reports, one highlighting in-group emphasis – which phrases were originally said by an individual but then promoted by the group – and another providing an automatically generated, anonymized summary of the meeting's broader points.

Another app I'm excited about is Polis, which has been used in Taiwan to extend direct democracy to citizens. Using the Polis app, participants can post comments about topics or proposed legislation; other users click 'agree', 'disagree' or 'pass/unsure' on each comment. Interestingly, participants can't respond directly to one another's comments, eliminating thread-derailing arguments and abuse. Polis was used in 2015 to set Taiwan's policy for ride-sharing apps like Uber, which faced opposition from local cab drivers. After a long series of posts to the app, the community came to a consensus that allowed Uber to operate in the country, but subjected it to strict regulations. Polis is deliberately designed to promote the norm, rather than extremes. It also uses AI to debunk disinformation before it gets a foothold, suspending and eventually banning participants who post it.

I am also enthusiastic about the ongoing research in this area by our friend Jonathan Zittrain, a professor at Harvard University. Zittrain grew up moderating CompuServe forums, and, after getting a law degree, co-founded Harvard's Berkman Klein Center, where Rosemary has been a fellow for the past dozen years. He recently launched the Institute for Rebooting Social Media at Harvard to assess and build social media technology in the public interest.

Jonathan believes that social media needs government assistance, in the manner of public broadcasting, to produce higher-quality content. 'It may be that you need a subsidy to produce and possibly to maintain something that isn't the McDonald's of social media,' he said recently. 'You need to subsidize something in the public interest. We need a kind of social media platform capable of winning a Pulitzer Prize.' Another idea of Jonathan's is to offer fine-tuned controls for the algorithms that feed content to us. 'It certainly would be nice if there was a way to turn the dial to something more positive,' he said.

One way to do this would be to craft algorithms specifically for vulnerable users. One for teenagers, for example, that steered them towards healthier content; perhaps another for people who seem to be falling into conspiratorial rabbit holes. If you have two social networks, one that's polarizing, and the other that's less addictive but more healthy, you could give your kids the healthy one. Or maybe you could yourself go into the controls and tell the app you're struggling to disengage; the algorithm would adjust to serve you less provocative content.

It's not enough merely to fix things. Many years ago, when I was working at Plessey in the UK, I was enrolled in a management course, where I was introduced to the 'Motivation-Hygiene' theory advanced by the business psychologist Frederick Herzberg. Despite (or perhaps owing to) its odd name, the motivation-hygiene concept has stuck with me over the years. Herzberg posited that the factors that lead to workplace dissatisfaction are completely separate from the factors that lead to workplace satisfaction. For example, if you have a workplace that's loud, filthy, noisy, etc., those things will hold you back – but it is not enough just to fix these issues. Once you've cleaned the place up, you need to focus on the motivation. You need to give your employees a mission to strive for, in order to create value in their lives. And the perfect time to begin to engineer this moment for the web was on a very special anniversary.

CHAPTER 14

The Contract for the Web

In 2019, the web turned thirty years old. On 12 March Rosemary and the Web Foundation team organized a whirlwind tour of four cities around the globe, commemorating the web's past, present and future. It was perhaps the busiest single twenty-four hours of our lives. We woke up in Geneva for a celebration at CERN, where the web was born. Much had changed at CERN in the intervening years. The 27-kilometre ring now housed the finished Large Hadron Collider, by far the most powerful particle accelerator the world has ever seen. One of its objectives was to search for the Higgs boson, a particle which interacts with other particles to give them mass. Without the Higgs, there would be no solid matter – no planets, no stars, just a universe of energy whizzing around.

The British physicist Peter Higgs had first proposed the existence of this particle in 1964, when he was thirty-five years old – about the same age I was when I had first proposed the web. Forty-eight years later and following significant investment, the existence of Higgs's theoretical boson was definitively demonstrated when a 14-metre-tall detector located in a cave on the French side of the border picked up traces of a never-before-seen particle that closely matched its profile. Higgs, then eighty-three, was in attendance at CERN's auditorium in 2012 when the discovery was announced. Upon learning of his achievement, he removed his glasses and wiped away tears: theory had intersected with practice, and his life's work was complete. The following year, Higgs was awarded the Nobel Prize in Physics.

That morning in 2019, following a reception at CERN's glass-box restaurant, I presented my own report on the state of the web at CERN's auditorium. Many old friends and colleagues were in attendance, including Ben Segal, who'd convinced CERN to adopt the global Internet Protocol standard so many years before, and Robert Cailliau, the first web evangelist, my colleague at CERN who'd been among the first to see the potential of the web. My public remarks that day closely tracked those of a letter we'd posted to the Web Foundation's site. 'The web has become a public square, a library, a doctor's office, a shop, a school, a design studio, an office, a cinema, a bank, and so much more,' I observed – but I also saw darker forces at work. 'While the web has created opportunity, given marginalized groups a voice, and made our daily lives easier, it has also created opportunity for scammers, given a voice to those who spread hatred, and made all kinds of crime easier to commit.'

With the web turning thirty (and me turning sixty-five) I wanted to recapture some of the initial optimism the technology had generated back in the 1990s. In this spirit, the Web Foundation team produced a new foundational text we called the Contract for the Web. Our inspiration was the Universal Declaration of Human Rights, which the United Nations (UN) adopted in 1948. Over time, several other declarations of this type had been adopted. For example, the UN Convention of the Law of the Sea, first adopted in 1958, attempted to collectively manage the world's oceans, while the Outer Space Treaty, passed in 1967, prohibited nuclear weapons in space and limited the use of the moon to peaceful purposes. Each treaty was an admirable attempt by the world's great powers to confront the difficult task of managing a collective resource that belonged more to humanity as a whole than to any specific country. And, of course, CERN, with its successful model of international cooperation, was the ideal place to present our Contract for the Web.

Announcing the Contract as part of the 12 March festivities generated a lot of buzz and support, with the analytics data showing a good amount of traffic to the Web Foundation website. Moments after presenting at CERN, we left to catch a flight to London, where 10 Downing Street had organized a reception to commemorate the web's anniversary. Following this, we travelled to London's famous Science Museum, where I was scheduled to give a public talk.

This was quite a spectacle. I had regularly visited the museum while growing up, and Rosemary had been there many times with her children. We walked through the museum's grand Victorian hall, passing James Watt's pioneering steam engine on the way to the event space, where perhaps a thousand people were waiting to see Dame Mary Archer present me with a Science Museum fellowship before my remarks around the birthday of the web. Tickets to the event had sold out within minutes. Those in attendance were, basically, fans of the web – they were my people. The Science

Museum's production team made the event appropriately colourful and dramatic. I felt like a rock star.

And yet I had barely any time to process it, as we had another international flight scheduled. Our itinerary for the day had been perfectly choreographed, but it required more or less everything to go smoothly. There had been a crash on the M4 motorway between London and Heathrow airport, which threatened our plans. Racing to the airport against traffic, we barely made our plane to Lagos, Nigeria, for the final events of the day – which was in fact the next morning but still less than twenty-four hours after we'd woken in Switzerland! There, Rosemary and I visited the Women's Technology Empowerment Centre, where we saw some of the work the young girls were doing, including learning to code in Python – here was the future of the web.

There is a constant need the world over to help women, and especially girls, get involved in tech – well, science, tech, engineering and maths – to make sure that all opportunities are open to everyone. And perhaps this is especially the case in developing countries to help them compete in the global marketplace. So it was important for us to make a stop there, to see the centre addressing this need. Nor were we finished! The following day, we got back on a plane to the UK, travelling to the small town of Kirkcaldy in Scotland, birthplace of the economist Adam Smith, to deliver two lectures alongside our friend Gordon Brown, the former UK prime minister, champion of open data and instigator of the Open Data Institute.

For me, the special thing about that day was that Rosemary gave a speech. I had given many, but she only a few. Addressing 'the

Economics of Inclusion', she talked about whether Adam Smith's invisible hand was serving the world. Three years before ChatGPT, she picked out AI as being a major future force of economic and societal change, warning of its huge potential impact on the world. She concluded that the invisible hand is not fit for purpose and, instead, we need an 'inclusive hand' that allows everyone to benefit from this transformative technology. The speech hit a chord with the locals, as a nearby factory had just closed. Rosemary's father had hailed from a nearby village in Scotland before emigrating to Canada in the 1950s, and I knew she was thinking of him as she addressed the room.

•

I had overbooked myself – but I felt it was important to do so. By 2019, the web was a worldwide phenomenon, as we'd hoped, and to limit our celebrations to just one or even two countries didn't honour its scope. We had seen a lot of discussions of digital rights, but what we needed was someone to say how we *get* these rights. We needed a global set of principles and commitments. That was what the Contract for the Web was for.

The idea of the Contract had originated with the senior staff at the Web Foundation. Adrian Lovett, the Web Foundation's CEO, convened a working group to draft the document. As with W3C, it was critical that a variety of different voices were brought to the table.

Following the announcement at CERN, and at other events, we presented the Contract in Lisbon, where, since 2009, the Irish entrepreneur Paddy Cosgrave has hosted the annual Web Summit conference, which is typically attended by tens of thousands of people, including a number of high-powered decision-makers from the tech and business worlds.

Ultimately, a core group of around twelve participants came together to draft the Contract. These included representatives from Google, Microsoft, the Web Foundation, the Wikimedia Foundation and the national governments of Germany and France. After several months of negotiation and debate, we settled on nine principles we thought the web should adhere to, with three principles apiece assigned to governments, companies and citizens. It turned out later that the chance to discuss the relative responsibilities of these different parties was something the working-group participants valued. The principles were broad, declarative statements that we endeavoured to make as easy as possible to understand.

FOR GOVERNMENTS

1. Ensure everyone can connect to the internet

Through my work at the Web Foundation, I'd come to understand in economic terms what being able 'to connect to the internet' actually meant, so it was possible to tie this principle to a specific goal: for citizens of Earth, 1GB of mobile data should cost no more than 2 per cent of average monthly income by 2025 – we called that '1 for 2'. I am very proud to say this has been achieved in the overwhelming majority of countries; for example, in Tanzania, 1GB of mobile data currently costs about 84 cents, which satisfies the requirement.

2. Keep all of the internet available, all of the time

Our specific concern here was government-triggered internet blackouts, particularly as a means of information control during periods of civil unrest. We've seen a lot of this over the years, notably in Iran, where the regime introduced a week-long internet blackout in 2019 in an attempt to suppress protests. Such broad internet shutdowns are a way of preventing citizens from organizing against authoritarian regimes.

3. Respect and protect people's fundamental online privacy and data rights

We have such an intimate relationship with our web browsers. Our deepest secrets are contained there, and digital citizens absolutely must have the expectation of protection from spying governments. Sadly, even Western governments routinely violate these rights, as the Snowden revelations show.

FOR COMPANIES

1. Make the internet affordable and accessible to everyone

This principle was drafted with low-income users in mind. Businesses naturally cater to the wealthy, as those are the customers with the most disposable income. We wanted a more equitable web. Among other goals, we wanted to ensure that the convention of network neutrality was respected so that internet service providers did not speed up or slow down web access based on users' choice of website.

2. Respect and protect people's privacy and personal data to build online trust

There is a feeling people get, based on the uncanny accuracy of targeted advertising, that your smartphone is secretly recording your conversations and sharing them with advertisers. The truth is even more nefarious – your phone doesn't *have* to record your conversations. The vast amount of data your phone harvests from you allows it to infer your interests, even as they develop! Still, the fact that so many people believe they are being illegally recorded suggests that companies have not done enough to safeguard privacy and data rights. This must change.

3. Develop technologies that support the best in humanity and challenge the worst

We wrote this in the hope that the web could remain a public good that puts people first. Investing in open protocols and building and developing the digital commons in an inclusive way is the key. The goal here is to promote connection between people, not engagement with algorithms. Most of what's on the internet is in fact benevolent – or at least not harmful – but too many people, especially kids, spend too much time using a small number of addictive social media services. In such a world, we must build better platforms and content to boost human flourishing and support democracy.

FOR CITIZENS

1. Be creators and collaborators on the web

The idea of intercreativity was at the heart of what I had always wanted the web to be – and of course, citizens had responded in exceptionally creative and intelligent ways. There are thousands of examples, but one is music; it is possible, using modern web technology, for musical collaborators around the world to write hit songs and even hit albums without ever occupying the same room. Here, the citizens of the world have exceeded all expectations.

2. Build strong communities that respect civil discourse and human dignity

The web can connect people around the world, and even people with different perspectives from within the same society. There will always be some friction when this happens, of course, but the systems we have now bring out the worst in us. The alternative is to build technical solutions that encourage people to find the common good in one another.

3. Fight for the web

This last principle was our way of reminding citizens that the open web we enjoy is not some eternal feature of our universe. There are constant political threats to the web's openness, and these must be combatted via political action, donations and, of course, the ballot box. Citizens must view the open web as a political right they fight for, just as they fight for freedom of assembly or freedom of speech.

Between my networking at the Lisbon Web Summit and various contacts at the Web Foundation, we were ultimately able to get about eighty major companies to sign up to the Web Contract, including Google, Microsoft, Amazon and Twitter. The contract was non-binding, of course, and there will always be bad actors who don't conform to some or even any of these principles. (That's true of the Declaration of Human Rights, too.) The idea is not so much to pass a law as it is to create a statement of principles and embed those principles, over the long term, into international norms. When we presented the Contract, some sceptics asked us what success would look like. We drafted the following answer:

> We will have succeeded when a critical mass of governments and companies have put the right laws, regulations and policies in place to create an open and empowering web for all; when they know their citizens and customers expect this of them; when it is the norm that most people communicate positively and respectfully online.

The Contract for the Web was a useful tool. The press could take it as a starting point, as could panel discussions. We made T-shirts and postcards of it. When people asked what they should be doing, we gave them the list. We pointed out that governments could not expect companies to change for the better unless they, the governments, also did their part.

•

How can this long-term vision become a reality? To start, I think, we must return to the decentralized structure I originally envisioned for the web. One could argue that the primary obstacle facing the web today is the power concentration among a handful of major sites. Along with search engines, Facebook, Instagram,

YouTube, TikTok and X handle an enormous amount of the traffic. As I've explained, each site is essentially a variation on the same concept, delivering media in an endless feed of algorithmically managed addictive content. The open space for collaboration and experimentation that defined the web through its first twenty years has been pushed aside by these new channels.

How has this happened? Instead of a tool for public good, the web has become more and more subject to capitalist forces tending towards monopolization. While Google, Apple and Meta are, at the time of writing, facing anti-trust lawsuits, governance, which should correct for this, has generally failed to do so. Regulatory measures have been outstripped by the rapid development of innovation, leading to a widening gap between technological advancements and effective oversight.

The working group that drafted the Contract for the Web brought technologists and governments together at the same conference table. This seems important – too often, the technologists are trying to guess what the government wants, and the government is trying to regulate technology that legislators don't fully understand. The evolution of networked computing has (somewhat ironically) led to a fracturing between the needs and interests of corporations, technologists, users and the government. We need common spaces for these groups to come together to discuss what the ideal future should look like. And we need new technologies to enhance human agency and provide the right incentives for corporations.

CHAPTER 15

Inrupt

We face a profound challenge. Polarization on the web and mental health problems arising from social media use must be addressed. Until they are fixed, it will be hard to get anyone to think about the web in a positive light. Until they are fixed, a lot of journalists, policy-makers, strategists and coders will struggle to imagine a better world.

There are already a multitude of good things on the internet, like email, calendars and podcasts. In each case, you have a free choice of app and each app is interoperable. For instance, email apps all work together – interoperate – because they all use the same email protocol, which is an open standard. Podcasts and calendars interoperate too. Social networks don't. Why can't I share my Facebook photos with my LinkedIn colleagues? Because Facebook, LinkedIn, Instagram, X, are all silos with no interoperability.

Accordingly, we need a new means of connecting and powering the social media we use. We need new open protocols and apps – a pro-democratic layer that empowers users and combats disinformation. Not only that, but to attract users, the new layer has to be much better than the old approach, unlocking fantastic new functionality as well.

There are many ways the web can evolve in these directions, but my hope is that the Solid protocol we have been developing offers one route. It can help us to accomplish three especially vital tasks: to break free from algorithmic manipulation, to unlock exciting new functionality and, ultimately, to transform the web user's digital footprint into an enduring source of value. The result could be a resurgence of optimism and faith in longstanding institutions.

Although we have HTML for documents, and URLs for making links, we have never settled on a standard for the *data* layer of the web. Instead, we let social media companies manage data for us, surveilling us and turning us into bundled products for advertising. This is what gives them the incentive to build these manipulative algorithms. The Solid protocol (solidproject.org) we'd been working on addresses these issues.

But how could Solid be built, and built to scale? To drive adoption of the Solid protocol, I needed somebody to build something for government and corporate customers as well as individuals. That is not the kind of thing that tends to happen for free. Only a well-capitalized, well-managed enterprise could attempt it. So in 2018, after years in the NGO space, I decided to do something I'd never done before. I decided to start a business.

•

Boston has a robust population of venture capitalists, and one who stood out to me was Rudina Seseri. Born in Albania, she had moved to the United States for high school and then attended Harvard

Business School. She later worked as an analyst in Microsoft's corporate acquisition group before striking off on her own as a venture capitalist. Her firm, Glasswing, had made a number of successful investments in the tech sector, with an emphasis on cybersecurity. She had a keen analytical mind and, like many of the best investors, the ability to digest information quickly and get to the point. Since 2014, I'd been having informal conversations with her about Solid. Our discussions had been mostly advisory in the early days but, from 2016, she too, began to see the need for something like the Solid layer on the web. 'What I was really looking for was not simply abstract talk about the value of privacy, but the idea that someone might lose revenue because of that,' Rudina said. 'You know, no privacy companies go viral, but Cambridge Analytica showed us how this could go really, really wrong.'

Rudina was also considering a (then) looming piece of EU legislation called the General Data Protection Regulation, or GDPR. The legislation introduced stringent requirements for how organizations collect, store and process personal data, making it essential that they have clear consent from individuals and granting the individuals rights to access, correct and delete their data. GDPR imposed hefty fines for non-compliance, creating a strong business incentive to implement better privacy controls. It also served as a model for similar legislation around the world, including in the UK, Turkey, Brazil, Japan, South Korea and the state of California. 'With the GDPR, enterprises realized that data privacy was something that *had* to be taken seriously,' Rudina recalled.

There is also the legal liability created by large-scale data breaches under current systems. When hackers targeted the credit bureau Equifax in 2017, they were able to obtain the personal information of more than 147 million people. This information was then resold to identity thieves on the so-called 'dark web', leading to all manner of credit-based scams. Equifax was ultimately forced to pay $425 million in restitution. Facebook suffered similar data breaches, and following a series of class-action lawsuits, its parent company Meta created a $725 million settlement fund. T-Mobile paid $300 million for a data breach; Capital One $190 million; Uber $148 million; and Home Depot $200 million.

At the time, Solid was an amorphous entity that Rosemary had been helping me formulate for years around the kitchen table. The protocol itself existed as an open specification published by my team at MIT and their collaborators. We had also convened an advisory board to think through a commercial approach – this eventually led to a new business called Inrupt, which combined 'innovation' with 'disruption', two of start-up land's favourite buzzwords. I had a vague mental picture of a logo: incoming green innovation disrupting the blue status quo. Inrupt.com was available, so I grabbed the domain name, then registered the corporation.

Initially, Inrupt wasn't much more than a legal entity. I needed a team, a staff, offices and funding. I also needed a CEO. As much as I fancied myself as a technologist – and as proud as I was of my work as the director of W3C – being the CEO of a corporation was not a role I felt well suited for. Fortunately, Rudina was able to connect me with John Bruce, an extremely sharp executive who'd previously managed one of her other cybersecurity investments, which had been recently sold to IBM. John was a savvy, quick-talking entrepreneur with an endearingly blunt manner and a no-nonsense management style. Like me, he was British and had brought up his kids in Lexington, Massachusetts. However, that was where our similarities ended. John spoke with a Liverpool accent, had attended the University of Bradford and had moved to the US early in his career while working his way up through the corporate world. In other words, he was both commercial and street smart, in contrast to my more academic manner. He became Inrupt's CEO, and I became the CTO. Together, we made a pretty good team. With Rudina's help, we raised multiple millions of dollars in our first investment round. 'When I'm in investments, a lot of them are ten, twelve, fifteen years,' she said. 'That's longer than some marriages.'

With so much riding on the integrity of the pods, the individual's secure store of personal data, Inrupt needed a world-class security expert to help secure the server protocol. Fortunately, John had worked with the famous cryptographer Bruce Schneier at a previous business. Like me, Bruce is a vocal critic of the way in which large companies have become central control points for our data, which he refers to as 'digital feudalism'. He grasped the idea for Solid very quickly, and his enthusiasm rivalled my own. 'The idea behind Solid is both simple and extraordinarily powerful,' he later wrote in a blog post:

> Your data lives in a pod that is controlled by you. Data generated by your things – your computer, your phone, your IoT [Internet of Things] whatever – is written to your pod. You authorize granular access to that pod to whoever you want for whatever reason you want. Your data is no longer in a bazillion

> places on the Internet, controlled by you-have-no-idea-who. It's yours. If you want your insurance company to have access to your fitness data, you grant it through your pod. If you want your friends to have access to your vacation photos, you grant it through your pod. If you want your thermostat to share data with your air conditioner, you give both of them access through your pod.

In 2019, Bruce became Inrupt's Chief Security Architect, heading up development of the suite of cryptographic tools that made this all function. It was a tremendous challenge, one befitting Bruce's reputation. 'Just trying to grasp what sort of granular permissions are required, and how the authentication flows might work, is mind-altering,' he wrote. 'We're stretching pretty much every Internet security protocol to its limits and beyond just setting this up.'

Inrupt and the Solid protocol was one of the more binary, moonshot bets in Glasswing's investment portfolio – as an ecosystem seeking a very broad base of users, either it was going to change the world, or it wasn't. 'When I do my portfolio construction, I take X amount and say, "I'm gonna swing for the fences,"' Rudina said. 'Make no mistake, this is one of those.' As part of Rudina's due diligence, Glasswing conducted a survey of the competition. There was barely anything out there. This meant my idea was either genius, harebrained, or simply so ambitious that no one thought they could pull it off. John liked it this way. 'I prefer businesses where nobody's ever heard of it before,' he said. 'They're more challenging, but they can create massive new markets.'

Glasswing's survey did identify a few companies in the

cryptocurrency space trying to do something a little similar, but none of these companies had made it very far. I was sceptical of blockchain-based technology, which the computer scientist and Ethereum co-founder Gavin Wood has grouped under the rubric Web3. This term, to me, was a meaningless buzzword as few of these initiatives had anything to do with the web per se (the moniker also conflicted with my own branding initiative for Solid, which I had been separately calling the 'Web – take 3.0'). My feeling was that the blockchain, the underlying technology that drove Bitcoin, was good for making transactions publicly identifiable, but a bad store for personal data. Not just because it's expensive and slow, but because any data you put on the blockchain is immediately public – not great for personal data, which should be private by default.

The other thing Glasswing's survey identified was a niche community of online hobbyists known as 'lifeloggers'. These enterprising individuals had essentially built their own homebrew pods, by downloading and saving their own financial and credit card records, uploading all their geolocation data, tracking the food they ate and their daily steps, requesting copies of their medical records and even their dental X-rays, and, in extreme cases, recording their daily lives with wearable microphones and cameras. Many people saw the lifeloggers as a little kooky, but to me they represented an interesting and important movement. Someday, I thought, everyone will live this way. If smartphones had represented an order-of-magnitude increase in data each person generated daily, I expected future developments in augmented reality and wearables technology to once again increase this data haul by several more orders of magnitude. Who would own this data? Facebook? To my thinking at least, this was personal, sovereign data, and it was a human rights issue that the user – the individual – retain complete control.

•

Inrupt rented an office building in downtown Boston, although our staff were located all over the world. We soon grew to about thirty people, but when Covid-19 arrived, we were all quarantined to our homes and conducted our business via teleconference. Fortunately, through W3C, I was already accustomed to remote work – in fact, we helped to build the infrastructure for it!

W3C had been working on in-browser real-time communication standards since the early 2010s. We published our first working draft of the Web Real-Time Communication (WebRTC) spec in 2011, and this powered the first cross-browser video call in 2013. Following a massive amount of work, W3C published the official standard in 2017 and has been evolving it ever since. Zoom, Google Meet and Microsoft Teams all use the WebRTC standard for large-scale teleconferencing – as does the free videoconferencing alternative Jitsi and a number of other open-source alternatives. Of course, all these services saw massive growth in usage during and after the pandemic. So did the open-source platforms I support, like Jitsi. It was enormously gratifying to me to see W3C's work help so many people around the world stay connected with colleagues and loved ones during Covid.

To be sure, it had taken ten years of development and a global pandemic for WebRTC to really take off. I was hoping Solid might have a smoother path to acceptance. John and I discussed a variety of strategies to build the Solid ecosystem. One was to offer users a personalized data dashboard that would allow them to see their entire digital footprint at a glance. *I* would have loved a product like this, and in fact that's what the Solid pod I'd customized for myself looked like. After some discussion, though, John convinced me that such a lifelogging platform put too much of a burden on the consumer. If you wanted to see your financial records, for example, you had to download them as a .csv file from your bank, clean them up into something useful, then upload them into your secure online pod. It was even harder to get things like medical records, which might involve multiple emails and phone calls to your health provider to be released. Even when this data was shared, as in the UK's National Health Service NHS app, it was still stored in some remote database to which you were granted conditional access. For Solid to succeed, we had to convince providers to generate the data using the Solid standard – and then import that data into an individual's pod, where they could access it via a dashboard.

People sometimes ask me what the 'killer app' for Solid is. I discourage them from using this term. There had never been a single killer app for the World Wide Web – instead, web protocols had opened thousands of uses, many of which I could never have imagined. If I had pigeonholed the early web and announced that the 'killer app' was a phone directory, or a database for academic pre-prints, which were among its first uses, I would have failed to communicate its huge potential. It's the same thing with Solid. You have a lot of apps on your phone, and you trust each app to access a certain slice of data. What if you could trust your apps to share data between themselves? They'd be much more effective this way!

At Inrupt, the plan is to build enterprise-grade servers for industrial users of Solid and see what develops. The first organization to contact Inrupt when we were in stealth mode was the UK's National Health Service in Manchester. They produced a proof-of-concept system that was able to write patient data directly to an individual patient's pod. In such a pod, we could then imagine combining that patient data with fitness-tracker data, from a Fitbit or a smartwatch – and with your consent, you could share the results with trusted parties. At present, no one has the ability to share any of their data in hospitals, and there are a lot of cases where it could benefit not just the patient but the doctor too.

Understandably, many people are reluctant to use genome-sequencing providers like 23andMe, as they don't want to donate their genetic information to a corporation which does not grant you privacy and control of your data. Incidentally, when 23andMe filed for bankruptcy in 2025, the question of who owned the data became more acute. But if your genome was stored in a Solid pod, it would be yours to own and inspect. With a simple click, you'd be able to share it with your doctor – if you wanted to. Think about what this could mean: if a company is distributing a new drug, and can anonymously and safely look at the genes of the people it has already given the drug to, it can start getting insights into how that drug works at a cellular level. And, of course, an individual's genetic information can talk to their fitness data to inform their workout routine. The benefits to prevention, treatment and medical research simply from letting this data reside in a secure, potentially shared space are tremendous.

Now let's add yet another data source: your spending habits. Credit and debit card data currently live inside inaccessible corporate databases, administered by your banks and the credit card companies. But there's so much potential value there for society and for individuals, if only they could get at it. Let's say medical researchers are trying to determine why a certain group of people get cancer. Give those people's genomes – and their credit card data – to a trusted AI to analyse. Suddenly, it would have a deep insight into their lifestyles: what they eat, where they go on vacation, how they spend their free time. It might lead to new research avenues or treatments. The patient, the doctors trying to help them and the health organizations all benefit. Or imagine you are looking to lose a few pounds. If you granted an AI access to your credit card data, and the data from your Fitbit, an AI could provide personalized tips. The possibilities are endless.

Governments accumulate enormous amounts of data on their citizens, but most of this data is stored in separate silos across dozens of agencies, sometimes on paper inside physical filing cabinets. For example, the US Internal Revenue Service has reams of information on every citizen, but to access it you have to file dozens of request forms. You often don't even know who holds what data about you. Things do not have to be this way. By combining income statements, property information, health care spending and personal financial records into a single personal pool, your taxes could be automatically and accurately done in a flash. Similarly, requests for permits, occupational licensing, military service records, green card and visa applications and all the rest could be accessed in a moment. Applications for new businesses, or educational grants, or construction permits, or your driving licence could be similarly made transparent and expedited. And this could all be possible with Solid's built-in controls to ensure both consent and a reliable record of who has access to what.

I suspect cultural norms would drive adoption of Solid in different ways in different countries. For example, when it comes to the culture of who people trust, there seems to be a big divide across the Atlantic, at least between the UK and the US. To make a huge overgeneralization, in the US, kids are taught in kindergarten not to trust the government, and to take up arms against it at a moment's notice. But Americans also have a touching faith in big business, to which they hand over all their personal data. By contrast, in the UK and the EU, people do tend to trust the government. There is a sense in which government is a chore, and they are grateful that someone else is doing it, so they don't have to. But they are very suspicious of huge corporations – typically American ones, of course. As a rule, when we have been approached by organizations in the UK and the EU, they have been governments or government agencies, and in the US they are big institutions, like banks and insurance companies.

Flanders, the Dutch-speaking part of Belgium, was an early adopter. The region had a lot of things going for it. It was small and agile, and it wanted data to be available as a national utility, along with water, gas and electricity. In the days when governments were learning to put their data online, the authorities in Flanders had made the region work much more smoothly by using common standards to share data between different departments. The minister-president, Jan Jambon, had worked at IBM, so he understood tech. In late 2020, Jan recorded a video telling the 6.5 million Flemish people that they should have control of their data, and that their interactions with government would henceforth be through their data pod.

Businesses would also benefit. The economic boost from the efficiency of Solid would, Jan proposed, boost Flanders' post-Covid recovery. For example, a company might request access to a person's delivery addresses through their pod. Rather than someone filling out their delivery address for every shop they bought from – having multiple entries of their address stored in separate silos – there would be one copy held in a pod that each shop could be authorized to access.

Flanders decided to deliver people data pods in a similar way to how they get broadband – through a utility company. There is a joint venture called Athumi, between the government and Inrupt, which actually runs the pods. Any one of the six million people in Flanders who interact with the government now has a Solid data pod, which they control, all run by Athumi.

•

When it took effect in 2018, the European Union's GDPR proved to be the toughest privacy and security legislation in the world, boosting the business case for Solid. Interestingly, when one of the drafters of the GDPR legislation was passing through MIT and saw a presentation on Solid, she was totally in favour of it. If those drafting the regulations could give consumers a set technology then they would give them Solid pods, she said. That was quite a motivator for the team. When you are designing something like the web or Solid it is always important to consider the technological and the political sides at the same time. You have to make sure that regulation and technology are a good fit.

GDPR itself offered a warning in this regard: it overregulated the delivery of cookies in the browser, leading to annoying 'cookie consent' pop-ups. The fines for non-compliance with GDPR were onerous, sometimes exceeding hundreds of millions of euros. As website operators sought to avoid gigantic international liabilities, consent pop-ups started to proliferate, even for users based in the UK and US, significantly degrading the browsing experience. Users often had no idea what they were being asked to sign – either they clicked 'yes' or 'no', or they simply closed the pop-up without taking any action at all.

This was unnecessary and could have been avoided if EU legislators had taken the time to talk to W3C, the tech companies and others about what they wanted to do. Without informed technical cooperation, the legislators failed to classify the different types of cookies they wanted to regulate. If they'd done so, you wouldn't have to constantly say OK to things which are in fact not very harmful.

The good news is that there is a movement, after almost thirty years, to phase-out third-party cookies entirely. By late 2024, Firefox and Safari no longer allowed invasive marketers to plant tracking devices in your browser and follow you all over the web, though Chrome hasn't yet followed suit. First-party cookies still remain: those stored by a website to enhance the user experience in future visits. As for the cookie consent pop-ups – well, at least in theory, offering only first-party cookies could reduce the potential liability for web hosts, so maybe website operators won't feel the need for those irritating pop-ups. That said, there is still some privacy risk even with first-party cookies – they might be shared with third parties. So, maybe the pop-ups won't be disappearing just yet after all.

•

Whether it's because of their role in enforcing regulation or in managing data itself, governments are important to Solid. Part of the frustration with democracy in the twenty-first century, I believe, is that governments have been too slow to meet the demands of digital citizens. Solid offers an effective way to do so. I get excited talking about lucid, intelligible, highly responsive government systems. And, once you have a system that works that way, you've built a layer of trust that can extend into medical, legal, academic, and commercial and ultimately personal data-generation systems.

I also believe that sharing this information in a smart way would liberate it. Why is your smartwatch writing your biological data to one invisible silo in one format? Why is your credit card writing your financial data to a second silo in a different format? Why are your YouTube comments, Reddit posts, Facebook updates and tweets all stored in different places? Why is the default expectation that you aren't supposed to be able to *look* at any of this stuff? *You* generate all this data – your actions, your choices, your body, your preferences, your decisions.

If you could gather all this data in one location, you would actually be able to understand better who you really were. Plus, you would be able to make new connections – link your heart rate log to your medical records, link your location data to your spending habits, link your meal plan to your grocery bill. Your life could improve, and you would regain control of your digital footprint. You would be a sovereign citizen of the universe of information. The immense potential of this will only grow as a new wave of technology approaches – one whose power will exceed all those that have come before.

CHAPTER 16

Intelligent Machines

Towards the end of 2022, Rosemary and I were invited to a summit at Ditchley Park, a grand old English country house near Oxford. The estate has hosted British royalty going back as far as Queen Elizabeth I, and Winston Churchill famously used the giant house as a refuge during the Second World War, and there made the case for US involvement in the war to President Franklin D. Roosevelt's emissary Harry Hopkins. In the 1960s, the house was donated to the non-profit Ditchley Foundation, which now manages the space as an influential think tank. James Arroyo is its executive director. We had been to a few workshops there on various timely topics and also found James's advice invaluable through changes of the various organizations. Our invitation was prompted by a new geopolitical concern: the public debut of the incredible ChatGPT AI system in late 2022.

I was amazed by ChatGPT the first time I used it, just as I'd been amazed by the RSA encryption algorithm and Google's PageRank. Each was a step change in functionality for the web, and for the world: in security, in discoverability, and now in intelligence. We had been watching AI become more important in different places, and we had, like many people, been hopeful about self-driving cars. (Rosemary made an early investment in Wayve, impressed by its Kiwi founder Alex Kendall, a super-bright Cambridge PhD.) Despite all that, the ability of the Large Language Models (LLMs) used by ChatGPT and other AIs to produce high-quality prose instantaneously was a complete shock. ChatGPT and its rivals sparked a revolution in computer science – not to mention a frenzied rally in the stock market. It was like being flung ten or twenty years into the future.

My amazement was broadly shared, and soon the Ditchley Foundation decided it might be a smart idea to host some of the UK's leading technologists, strategists and politicians to hash out the implications. It was an honour to be among those invited.

There are twenty-six guest rooms at Ditchley, and when you arrive there to stay for a night, the staff ask if it's your first time there. If so, they inform you to be very careful in filling the free-standing clawfoot bathtubs in your bathroom. The plumbing system at Ditchley is ancient, but very powerful, and many an unsuspecting guest has turned on the hot water and walked away, only to return to a flooded room.

The conference rooms at Ditchley are of similar vintage, with a grand hall for breakfast and many large sitting rooms lined with leather-bound books. But Ditchley is not as old-fashioned as it appears. In the large library where we met, many of the bookshelves concealed cameras, and the table contained a host of modern teleconferencing equipment, as well as a huge screen. As the meeting began, AV technicians located in an adjacent room began stitching together a well-edited video digest of the discussion in real time. The room felt almost like something a Bond villain would create – a fascinating mash-up of an old-school British grand strategy conference with a modern curated panel discussion. Ditchley rules prevent me from quoting anyone directly, so I must be a little circumspect in what I describe. What I can say is that it was clear to everyone at the conference that a new era of technology had arrived. The people working on AI itself were typically as surprised as the rest.

I have always felt it is important to have an AI that works for *me*, and not for big tech. We already have strict cultural expectations that my doctor works for me and must always have my best interests at heart. My lawyer, too, must always work on my behalf. We enshrine that place in our culture with legislation, regulations, special confidentiality privileges and high professional standards for these trusted roles. Could one not make the same argument – and potentially adopt the same kind of regulations – for AI? A breakout session at Ditchley gave me the chance to try out this idea on the rich mix of scientists, politicians and philosophers sitting around the table. They seemed to think it made sense, was valuable, and that it could be done. That has been a good reference for me going forward.

•

In order to better understand AI's promise and peril, and what I had to say about them at Ditchley, a little history is in order. As I've mentioned before, the AI world had long been divided into two: the 'logic' camp and the 'neural net' camp. The neural nets, which mimicked the structure of biological brains, had initially been deemed a failure. But, as the hardware on which they ran became more powerful, they started to produce extraordinary results. They were recognizing faces, beating world-class players at Go and producing high-quality synthesized speech. In the late 2010s, a research team at Google developed a powerful new neural net architecture termed a 'transformer'. Following this, researchers at OpenAI, which was then a non-profit start-up, adopted Google's transformer architecture. They began training it on large quantities of data, using it to generate new text. They called this tool a Generative Pre-trained Transformer, or GPT.

The earliest versions of GPT were not very impressive – partly because the data sets OpenAI researchers were using weren't large enough. GPT-1, for example, trained on a corpus of self-published e-books, mostly from the romance and science fiction categories, and its output was closer to nonsense than meaningful human prose. It was only once OpenAI researchers started training their data on the web that they began to see meaningful results.

To do so, OpenAI researchers turned to the non-profit archivists at Common Crawl. Founded in 2008, Common Crawl was an alternative to the Internet Archive, geared towards academic researchers. Its founder, Gil Elbaz, was a successful Silicon Valley entrepreneur. Like me, he'd been interested in semantic data structures, and in 2003 he sold his start-up Applied Semantics to Google for more than $100 million. This ended up being a fantastic deal, at least for Google; Applied Semantics provided the core of Google's AdSense product, which even today remains their engine of profitability. I have never personally met Gil, but I am very impressed with his non-profit work. Every month, Common Crawl makes a complete backup copy of the entire web. The Common Crawl data set archives about 50 billion web pages, and runs to around 7 petabytes; with 8 billion people on Earth, this works out to about 1 megabyte per person.

The success of large language models like GPT was due to their ability to scale – and, when it came to language, the web was the single largest data set there was. So, by training a transformer against the entire web, OpenAI built the most powerful large language models anyone had ever seen. GPT-3 and its successor models were astonishing tools that shocked not just the public but even experienced AI researchers.

Although at some level you might say the GPT models were 'merely' crunching zeros and ones, there seemed to be intelligence within the machine. People were (and are) very divided on this. Some believe that because the model has been trained simply by trying to guess the next word in a sentence, it can never really be intelligent in the way we are with our brains. I think these sceptics fail to understand the effect of scale. Yes, an LLM is about guessing the next word, but when you feed it a high proportion of all the sentences in the whole world, it builds a network with the number of nodes comparable to the number of neurons in your brain.

People have been arguing about this topic for some time. Back in 2010, I attended a lunch at the Google Zeitgeist networking event, and ended up talking with Google co-founder Larry Page. Even then, AI was beginning to achieve much of what my parents had told me computers couldn't do, like translation between languages, recognizing objects visually, and planning a route across a city. AI folk would complain that every time they got a new thing working, people would stop calling it AI. Instead, they would just call it 'language translation' or 'pattern recognition', or 'satnav', so AI would not get the credit. When I asked Larry what he was interested in, he said, 'AI'. I mentioned to him that specialized AI could mimic many of the systems in the brain, but that there did not yet seem to be an AI implementation of a 'stream of consciousness'. Then, I wondered aloud whether what we called 'consciousness' was just another system of the brain, one that, like language translation and pattern recognition, would not seem to be such a big deal once we had found it. 'You write the code,' Larry replied, 'I'll run it.' In fact, as Google ran perhaps the largest private collection of computers in the world even then, Larry reckoned he could put together a neural net *bigger* than the brain if needed.

The computing power available to Google has increased by orders of magnitude since that conversation, so in a sense I have always been expecting someone to write that code for the stream of consciousness, and someone somewhere to run it. In fact, we may have already met that threshold. Philosophers and scientists hold many different views on the subject, but in my own, there is no difference between simulating consciousness and having it; any intelligence is what it appears to be. While ChatGPT is not wired up to present a stream of consciousness, my guess is that it may even be smart enough to do so if it were wired up differently.

My take has always been that there is no reason to distinguish between silicon and organic material – intelligence isn't a function of what you're made of. Alan Turing saw this too, which is why his proposed test for the presence of intelligence was conducted from behind a screen. In 1950, he'd speculated that the technology required for a machine to pass his test might arrive by the year 2000. It took longer than that, but not – in the scale of history, or even of our lives – all that much longer. In 2024, researchers at Stanford conducted a rigorous version of the Imitation Game using OpenAI's latest GPT model. The chatbot resoundingly passed the Turing Test – its output was statistically indistinguishable from tens of thousands of human subjects from more than fifty countries. On certain questions, the AI was even 'better' than humans, showing more evidence of compassion and trust, and even incipient signs of personality and mood. Data was the fuel for this breakthrough – free, open web data in particular. From the Ferranti punch tape strewn around my parents' house when I was a child, we have arrived at today's evolving, intelligent machines.

•

At Ditchley, the discussion centred on the implications of this unprecedented breakthrough. For years, technologists had discussed the implications of what might happen if humans invented a system that was smarter than ourselves. The futurist Ray Kurzweil, who, like Isaac Asimov, has been thinking about this for a long time, popularized the notion of the 'singularity', which occurs when AI can build its *own* AI, triggering an uncontrollable explosion in intelligence with unforeseeable consequences. Ray thought that the advent of superhuman intelligence would be something we had to be very careful about, and many researchers agreed. Now the question is forced upon us. For most of my career, there were people who believed that computers would not be smarter than us for a very long time, and even people who thought that computers would *never* be smarter than us. After the debut of ChatGPT, these positions became much less tenable! People might argue about just how far we are away from the singularity. But if they argue it is impossible, it is from philosophical or religious standpoints rather than the practicalities of it. Speaking with the attendees at the Ditchley workshop, it was clear that many people's timelines for the arrival of the singularity had collapsed.

Today, experts in the AI field are rather crudely characterized as 'AI boomers' or 'AI doomers'. In my opinion, it is not a useful dichotomy: it tries to divide the world into people of extreme opinions. Maybe the polarizing social media, maybe an impatient press have made this happen. Writing this book now, I realize that those words are unhelpful, in the same way that it's not helpful when people ask, 'Is regulation good or bad?' It really depends on what you are thinking of regulating! The potential impact of AI is complex: your answer must depend on context. Are you looking at the future of, say, jobs in the legal industry, or the discovery of new cancer drugs, or the containment of superintelligence. You can't lump it all together! Are there any pure doomers who only see the bad sides? Not that I have found. Are there any pure boomers, who only think of the good? Well, maybe one.

Marc Andreessen – yes, the coder at the NCSA who made the Mosaic web browser and co-founded Netscape back in the day – now co-manages the Andreessen Horowitz venture-capital fund in Silicon Valley. He is a self-styled techno-optimist, and his June 2023 blog 'Why AI Will Save the World' makes fun of anyone who has any concerns about AI as being scared of 'killer robots':

> My view is that the idea that AI will decide to literally kill humanity is a profound category error. AI is not a living being . . . It is math – code – computers, built by people, owned by people, used by people, controlled by people. The idea that it will at some point develop a mind of its own and decide that it has motivations that lead it to try to kill us is a superstitious handwave.
>
> In short, AI doesn't *want*, it doesn't have *goals*, it doesn't want to *kill you*, because it's not *alive*. And AI is a machine – is not going to come alive any more than your toaster will.

Marc clearly can't imagine something in the cloud or on his computer being smarter than him and overtaking him. Apart from Marc, I think most other people can see the complexity of AI and the complexity of its impact.

I'm not terrified of AI. But I think we do have to control it. In the movie *Ex Machina*, the intelligence not only has to be smarter than a human being, it also has to be in the physical form of a human being – a beautiful blonde female human being – to escape its captors. In reality, AI doesn't need a 'robotic' physical presence to be dangerous. Even as a cloud presence, a superintelligent AI could influence opinions, manipulate stock markets or, potentially, do even worse.

•

ChatGPT is a portal to OpenAI's latest models, and so the systems you interact with through the chat window are constantly evolving. At the time of writing, the most recent innovations are what we call

'AI agents'. We call an AI 'agentic' if it is autonomously capable of solving complex, multi-step problems. AI scientists have periodically rolled out such 'autonomous agents' over the years, but in the past they were always faking it. Once the fakery was discovered, the program was usually a disaster – remember Clippy, who wanted to chat with you about the letter you were writing in Microsoft Word?

Clippy would be very powerful now. Perhaps it was the memory of Clippy that prompted the philosopher Nick Bostrom's hypothetical 'paperclip maximizer', the idea of an AI that, like the brooms in *The Sorcerer's Apprentice*, fulfils its directive to make as many paper clips as possible by transforming all atoms (and in the process, all humans) into paper clips and thus annihilating the universe. This argument drives me crazy, because this hypothetical hypersmart paperclip AI is really very *dumb*, with zero embedded controls. Nothing we're building resembles a paperclip maximizer – in fact, responsive systems like ChatGPT are already far smarter than it. ChatGPT would know it's making too many paper clips. It would know because you told it so.

But we are approaching a near future in which we will ask AI to autonomously pursue goals. And if you start to give AI agents goals, those goals might not be aligned with humanity's best interests. We should work very hard to make sure we stick with systems that we can contain. While we don't have paperclip maximizers, we do have *profit* maximizers: we call them corporations. They often prove hard to control! An AI placed in charge of a corporation and tasked with maximizing profits might do many things we cannot expect. Not only that, but in America, a corporation is granted many of the same legal rights as a person. Imagine an AI running a fossil fuel company, with AI PR people and lobbyists. You can see how this would be a bad idea.

One technique to control a powerful AI is to guarantee that advanced systems are always programmed to return to humans for approval. (This is sometimes called the 'human in the loop' doctrine.) But how is the human going to validate what the AI has done? It may be easy to reject a plan to convert the universe into paper clips, but it's hard to override a complex medical diagnosis.

If we are about to make something smarter than ourselves, the wise thing to do would be to build it in a sandbox where it can play, but not affect the real world – where it does not have the power to argue for its own improvement, or to be given more resources. To ensure compliance, we require global collaboration and governance – a sort of CERN-like institution where the very best AI researchers can come to collaborate, rather than compete in separate companies as they do now. Indeed, in late 2024 Demis Hassabis, the co-founder of DeepMind with Mustafa Suleyman, issued a call for just such an institution. He is not the only one. One of the benefits of being the inventor of the web is that I am included in a number of global conversations cutting across disciplines and sectors. We need more of these conversations to drive AI collaboration forward. A CERN-type institution could help to bring people together from the academic, governmental, philanthropic and commercial worlds.

CERN's history can inform us here: it was founded just a few years after the appalling bombings of Hiroshima and Nagasaki. As CERN provided a collaborative environment for research, the Strategic Arms Limitation Treaty and other similar bodies managed the

Cold War standoff without destroying the world. That was the time into which I was born, when nuclear annihilation was our chief fear. In the modern era, we have climate change to worry about along with AI, although of course the threat of nuclear weapons has not gone away. For international collaboration on climate, there is the UN and its yearly Conference of the Parties (COP). We are seeing the embryonic signs of similar initiatives for AI.

In 2023, the UK convened the first global summit on artificial intelligence, held at Bletchley Park, fittingly where Alan Turing used to work as a codebreaker and computing pioneer. The AI Safety summit drew a number of prominent attendees, including Rishi Sunak, then British prime minister, US Vice President Kamala Harris, Elon Musk and Wu Zhaohui, Chinese vice-minister of science and technology. Subsequent AI summits in Seoul (2024) and Paris (2025) have been equally well attended but have focused more on innovation. The Paris conference was tellingly named 'The Artificial Intelligence Action Summit' and co-chaired by French President Emmanuel Macron and Indian Prime Minister Narendra Modi. This shift from Safety to Action has meant less coordinated work on AI guardrails. And with the competitive momentum behind the various LLM projects across the world, it is a challenge to balance different commercial interests. That is precisely why we need to double down on international dialogue. There is still the opportunity to establish a CERN-like institution for AI *before* something catastrophic happens.

•

AI offers tremendous utility. I use several AI tools myself – though the tools available and my use of them are constantly changing. One I've made part of my workflow at the moment is Microsoft's Copilot, an AI assistant which is pretty good at writing simple computer code or helping with articles, like my Design Issues blog posts. As time goes on, I find myself delegating a certain amount of the grunt work to Copilot, including public code snippets I've shared on my blog and elsewhere. Of course, automating the grunt work means I spend more time stuck on hard problems! Here, too, though, AI is useful, especially ChatGPT. When I'm really stuck, I typically give ChatGPT a pointer to the software repository I'm working from and describe the problem I'm encountering. There have been times when it has proven invaluable – especially when a really obscure error message pops up, and it can figure out the sequence of errors which led to it.

As time goes on, my programming environment will grow more integrated with AI. This will result in an ironic, full-circle revolution. When I started out, almost all programming was done via the command line, by entering text with a blinking cursor. The user typed a command, and the system replied with more text. When the NeXT debuted, with its elegant graphical user interface, programming was transformed into a more drag-and-drop affair. But with AI, we have another paradigm shift, returning us to the command line! Only this time the command line is a *conversation* line, hugely more capable and intelligent, and not requiring specialized vocabulary to control.

Just as it is transforming programming, AI is going to transform the web-surfing experience. Already, perhaps 30 per cent of new searches involve the use of an AI engine. For the younger generation, to 'chat' something is becoming a verb, in the same way we might 'google' something. As GPT models train on ever-larger amounts of data, they are growing very smart – OpenAI's latest GPT model can produce sophisticated 'chain-of-thought' mathematical reasoning, and in a recent (small) study, outperformed doctors in giving a medical diagnosis. What a great accomplishment this is.

Like many other people, I've also started using AI for more personal queries. It's quite fun to interact with these systems, and they often have surprising insights. With all my focus on an AI that works for *me*, I was happy to see Inflection.AI, a company co-founded by Mustafa Suleyman, which respected the privacy of the individual by integrating sensitive user data but not extracting it. Inflection's values seemed aligned with Inrupt's, and so I called Mustafa out of the blue to pitch him the idea of integrating his AI with Solid. Unfortunately (well for me, anyway), in 2024, Mustafa and most of the rest of his team were tempted away from Inflection for a large sum by Microsoft. But I still use Inflection's confidential agent, called pi.ai, for all kinds of purposes. Of course, if I could give it access to all my personal data, it could maybe do much better.

The move from web surfing to user-led conversations seems to me to be a healthy one. The current flow tends to be: Google for what you want; follow links to find what you want; get distracted by ads to start looking at something else; buy something. The AI-based flow is more like: express what you want; refine it in discussion with an agent which understands you and the world out there; and select a very good fit for what you want. I'll talk more about this switch from the attention economy to the intention economy in the next chapter.

To get to that point, the problem of data sovereignty is one of many looming issues with AI that we'll have to get right. Many of the most powerful new AI models are Retrieval-Augmented Generation systems, or RAGs. These systems can synthesize public web data with private enterprise data; or private *individual* data, such as what's on your phone. Data remains the fuel for AI, and if we can't develop systems with user control at their core, I fear we will encounter the same problems with privacy and exploitation as we have with social media and search. It's early days, but as these companies move to monetize revenue, there is an ever more pressing need to protect our data.

Of course, AI as currently practised has many problems. One of them is that falsehood and malice have led to the existence of material on the web which is just wrong, and it is all too easy to include erroneous information in the data your AI is trained on. There's an adage in computer science that dates from the 1950s: 'Garbage in, garbage out'. The saying remains as true today as it was in the days of the IBM mainframe. An LLM trained on inaccurate information, on hate speech, or on deliberate disinformation, will reproduce those same flaws in its results.

To prevent the nastiest outputs, LLMs like ChatGPT are fine-tuned using 'reinforcement learning from human feedback', an AI technique that incorporates editorial judgement from humans. Of course, this immediately leads to the objection that the output of the model is being censored. Conservative commentators were quick to decry some of GPT's limitations and were furious when Gemini, Google's LLM, retold events from American history through a filter of what they saw as excessive political correctness. The truth is that these systems *need* a filter of some kind; and with dozens of firms racing to build LLMs, I suspect each system will change massively in this respect.

Another problem is that LLMs train on a vast database of copyrighted data. The Common Crawl web corpus is 'free' in the sense that it does not cost money to download, but it contains a lot of copyrighted material. Overwhelmingly, the data set of Common Crawl contains words that people wrote, that belong to them – that are their intellectual property. LLMs can reproduce this text, sometimes subtly, sometimes even verbatim. This unfairly exploits creators and generates significant legal liability for the LLMs' builders.

In December 2023, The New York Times Company sued OpenAI and Microsoft for copyright infringement. The *Times* alleged that millions of its articles had been used as training inputs for the GPT models and that, with a little prompting, GPT users could get the LLM to reproduce segments from those articles word for word. (The *Times* cited as an example critic Pete Wells's notorious takedown of Guy Fieri's midtown tourist trap restaurant. When prompted, GPT copied Wells's language exactly: 'Why did the toasted marshmallow taste like fish?') Another, similar suit was filed against OpenAI by a coalition of bestselling authors, including George R. R. Martin and John Grisham.

It's unlikely this will be resolved quickly, and the fight for attribution and fair compensation spans all creative industries and countries. In 2025, more than 1,000 music artists, including Kate Bush, released a silent album to protest against the UK government, which had proposed to allow AI companies to train on all creative content unless their creators had explicitly opted out. On the last day of the government's consultation over these proposals, the 'Make It Fair' campaign was supported with front-page coverage on every major UK newspaper.

An AI should only have the right to use your creative work as input if you have explicitly opted in, and should you opt in, you have a right to get paid. There's a template for this. Years ago, YouTube faced a similar problem regarding the unauthorized use of copyright music in uploaded videos. In response, YouTube built its Content ID system, which scans videos and compares them against a database of recorded work. When it finds a match, it gives the rights holder the option of either muting the music, pulling the video entirely, or directing a share of the video's revenue streams back to the rights holder. Some amateur content creators have objected to this 'demonetization', but ultimately Content ID is a way for the artist to get paid.

We now need to build a similar identification system for AI. Unfortunately, this is a much harder technical problem. Determining how much each individual sentence contributes to an LLM's output is explored in the nascent discipline of 'AI attribution'. In some cases, like that of Pete Wells's review, the attribution is 100 per cent – but that is rare. Let's take a more challenging case. Here, I asked GPT-4 to write me an ode to the dishwasher, in the style of the great English poet Philip Larkin:

ODE TO THE DISHWASHER
In this quiet kitchen corner, humming low,
It takes the mundane, the daily grime and grind,
The plates and cups, the remnants of our show,
To cleanse, to rinse, to leave no trace behind.

No pomp, no grand applause, it toils away,
A servant to our careless, cluttered lives,
With each cycle, it erases the display
Of dinners past, where conversation thrives.

Yet in its hum, a melancholy sings,
Of routines, of the roles we all must play,
Of fleeting joys, the little, simple things,
The cups we fill and drain, day after day.

Not bad. Not bad at all. Astonishing, really, for a computer, and it took GPT-4 about three seconds to produce. But how much should Larkin's estate get paid for this? Larkin never actually wrote an ode to the dishwasher, and no single line of this poem directly plagiarizes from anything he ever wrote. Yet the style is inimitably his! Trying to determine how much this poem 'plagiarizes' from Larkin, if at all, is not an easy question.

The Common Crawl corpus was a large data set, certainly, but it was a small fraction of all the data online. The size of the internet's total data set is measured in zettabytes, each of which is the equivalent of a trillion gigabytes. Of course, a lot of that data isn't easily accessible. It's stored behind password-protected firewalls, or on websites that don't give search engines permission to archive them. This 'hidden' internet is sometimes referred to as the 'deep web' and it's estimated to be several times larger than the public-facing portion.

As you read this, AI researchers are burrowing into these vast troves of data to build ever more powerful systems. Sometimes they may delve too far; in June 2024, *Business Insider* reported that both OpenAI and Anthropic were ignoring the web's robots.txt convention to scrape data from sites that had opted out (although both companies have publicly maintained that their web crawlers do indeed respect robots.txt). Part of the logic of Elon Musk's acquisition of Twitter may have been to use the site's vast archives of conversational, human-generated text to train his Grok AI system. Google's system Gemini could train AIs on the company's YouTube database and Gmail archives, as well as the vast, private trove of queries that users have punched into the company's search box since 1998. Meta can train AIs on Instagram and Facebook data – you may have unwittingly signed your rights to do so over to them years ago, when agreeing to the sites' terms of service. And ByteDance can use TikTok, whether or not the service remains active in the US; in fact, ByteDance was one of the largest buyers of Nvidia's AI-training microchips before the Biden administration blocked the further sales of such hardware to China.

This leads us to another problem: the rise of the deepfake. Deepfakes are highly realistic synthetic images, audio and video that can be repurposed in all sorts of nefarious ways. One common scam features a distressed call from a relative who is supposedly being held hostage. The scammers then take the phone and demand immediate payment – before the victim realizes that their loved one's voice has merely been scraped and faked. More elaborate deepfake frauds are possible using teleconferencing. In Hong Kong in 2023, a finance worker was tricked into paying out $25 million after fraudsters impersonated the company's chief financial officer in a video conference call. Several of the victim's co-workers appeared to be present in the call; all of them were deepfakes as well.

Deepfakes pose a threat to informed democracy. They can be used to make political candidates seem to say all manner of outlandish statements. But they can also be used in more subtle ways, like telling voters the wrong day to go to the polls. Deepfakes can even be used to conduct voter outreach. In the 2024 parliamentary elections in India, campaign strategists scanned the voices of popular politicians, then used deepfake technology to craft targeted messages to individual voters. Imagine getting a phone call from your favourite politician, speaking directly on the issues you care about most! That technology is here, today. Terrifying, but also effective.

What scares me is the combination of deepfakes with microtargeted digital advertising. The problem with narrowcast delivery, as I said before, is that it's very difficult to debunk misinformation when it's only being shown to the most susceptible audience. Combining AI-generated deepfakes with AI-guided microtargeting could create a severely distorted perception of reality, almost a hall of mirrors. People might begin to doubt all media that they see. To a certain extent, this is already happening.

To help users and systems distinguish what is fake from what is real, there are two parallel ways to go – labelling the fakes and labelling the reality. If you label the fakes, that creates an incentive to remove the label. But when you create a label on a genuine image, there's no incentive to remove that digital signature from it. The labels are also encoded cryptographically so they are impossible to forge. Labelling the real things is called tracking their provenance. This involves recording the whole path, from a photograph being taken, through its being cropped and adjusted, then to its being included in a web page.

A lot of metadata is created when you take a digital photograph. Cameras and phones typically store the time and place, and photo-editing programs record that they have been used in their output. The Sony Alpha camera can make a 'digital birth certificate' when you take a picture, a cryptographic claim that you took exactly those pixels. If anyone tries to use that image with any changes at all, it won't work. There is already a standards body for metadata in media, the Coalition for Content Provenance and Authenticity (C2PA), which includes Adobe, the BBC and Sony, among others.

We need to harden this information by adding a digital signature that's cryptographically generated and shows that the image and the provenance could *only* have been generated by a given photographer's camera. The Bellingcat organization, started by Eliot Higgins, investigates controversial images – of wars, for example – and figures out whether they should be believed or not. Circumstantial evidence around a report can make this easier. If you really want to make an image safe, what you can do is embed not just the time and location data into it but also ask the camera to list all the wi-fi signals it noticed at the time. This information is hard to spoof and can be inspected by authentication specialists.

One thing I've long thought browsers should have is an 'Oh yeah?' button. You click on it, and the browser scans all image and video sources and shows the provenance of the information. For example, it might read, 'This image was originally captured by a photographer Alice at the Associated Press. It had its levels adjusted in Photoshop by Bob, and was then signed as authentic by the photo desk at organization C,' alongside a list of timestamps. Cryptographically, it's possible not just to put this data in the image timestamp, but actually to embed it in every stage of the photograph's life, establishing a sort of chain of custody.

We also need to come up with a solution for voice deepfakes. Like the RSA scheme we used in the past, we need a public and private key when logging into teleconferencing. If you don't enter your passkey or password when logging into the conference, a question mark should appear next to your name. Sadly, we can't continue to use the RSA prime-number trick for ever, as it is theoretically vulnerable to an attack from a sufficiently powerful quantum computer, should one ever be built. For this reason, cryptographers are preparing to move away from the RSA algorithm to other public-key-based authentication systems.

In fact, passwords in general are feeling a little stale, and a great many have been leaked and are now floating around in the dark web. Because setting up a new password is a pain, people tend to reuse old ones, even when they're vulnerable! Google is now (gently) pushing for passkeys to replace passwords. These passkeys are cryptographic keys stored on your device, typically a laptop or phone. You access them via biometric identification – either your face or a fingerprint. To make sure you really are who you say you are in that video call, you may soon have to pass a biometric scan on your local device.

•

We discussed all these problems to various degrees at Ditchley – but I want to return for a moment to the question of superintelligence. All the conferences which talk about AI without distinguishing this issue from the others leave the world with a muddled set of priorities. Superintelligent systems might have goals that are incompatible with human well-being and survival. This, the 'alignment problem', is the biggest one we face. A misaligned AI might manipulate us or coerce us in ways we wouldn't necessarily perceive or expect. In extreme cases, a highly capable but misaligned AI could pose existential risks, potentially acting in ways that threaten humanity's survival.

Now, there are many in the AI community who find such speculations far-fetched. We've heard from Marc Andreessen. For his part, machine-learning pioneer Andrew Ng has compared discussion of AI risk to worrying about overpopulation on Mars. But there are others, including some of the world's most eminent computer scientists, who are beginning to worry. Of the three men awarded the Turing Prize for AI research in 2018, only Yann LeCun is comfortable with the risks. His co-winners, Yoshua Bengio and Geoffrey Hinton, are wary. They are also currently the two most cited computer scientists of all time.

I think we should listen to Yoshua and Geoffrey. Sci-fi (and particularly Isaac Asimov) has trained us to think that we can simply give AI rules, like 'don't kill a human being', and it will obey. But the neural net systems don't work that way; they aren't rules-based, and the complexity of their behaviour will soon approach our own. This is a hard problem, arguably the hardest and most important in computer science. No one yet has a solution.

But it's equally a mistake to focus *only* on the risks, when there is so much potential for these systems. Suppose we *could* definitively solve the alignment problem, guaranteeing that AI will always act according to human interest. And suppose we really trusted that AI, in the way we trusted our lawyers or doctors. In the next chapter, let's consider what that world might look like.

CHAPTER 17

Attention vs. Intention

In 2022, Rosemary and I travelled to Seoul. We had visited South Korea before – we had even dressed in robes for a tea ceremony – so we had some idea of what to expect. This time, though, the trip was to receive the Seoul Peace Prize, which I had been awarded. This human-centred award was a welcome surprise.

Wanting to learn more about the country, we arranged to have dinner with the UK ambassador on our arrival at a traditional Korean restaurant in the Old Town. What we hadn't realized was that Google Maps had been denied certain mapping data in Korea. It isn't until you end up in the dark, in the wrong area, on the wrong street, in a different time zone, in an unfamiliar country where you don't speak the language that you recognize quite how reliant you are on your usual digital tools. There was a mild panic and many hand signals to passers-by, who responded only with, 'Naver! Naver!' As the sky opened and a torrential downpour began, we found our way via Naver Maps to the restaurant. Arriving drenched, we reflected on how, after thirty years, the web had become an indispensable utility, as necessary as electricity. Navigating the modern world without it is almost impossible.

The South Korean economy was booming, and when I entered the award ceremony dinner the following day, I was greeted by a fairly intimidating-looking panel of leading scientists and industrialists. I was used to giving acceptance speeches, but this felt different. This time, the award was for 'promoting data sovereignty', and it specifically gave credit and honour to the open-source Solid protocol I was trying to get the world to adopt. I was deeply honoured, and the fact that it was a 'peace prize' made it all the more special. Peace is more than just the absence of war. It is, as the American author E. B. White said, 'The state of Something Good Happening'. As I'd tried to show with my Design Issues map of the internet, our aim should be not just to identify and fight bad things online, but to build better ones.

Data sovereignty is a term most people associate with nations, but in my speech, I suggested what it really meant was that *individuals* should have control over their own data. And not only control: personal data sovereignty should also allow individuals to be *empowered* by their own digital footprint. Going back to the example from years before, when I went skiing with Ben and we combined our GPS data on our computer with our photos, in a way we were empowering ourselves. I wanted that for everyone. And not just for GPS data, but for all sorts of data – when you link your health data with your financial data, and gain new insights, you are exercising your data sovereignty.

Receiving the peace prize brought home that people across the world were seeing the power of data – if only it could be ethically

Receiving the Seoul Peace Prize in 2022 – my first prize that recognized data sovereignty.

harnessed in a human-centric way. A few days after the ceremony, I met with the MyData folk in their offices in downtown Seoul. MyData is a global community that is pushing for individuals to have control of their own data, across a number of sectors. They have a local hub in Korea and it was pretty neat to talk with people who were fighting for the same global shift in priorities as myself. I felt energized by a sense of progress – wherever I went, people agreed that our relationship with online technology had to change.

•

The idea of the 'intention economy' had been circulating in my mind long before I first discussed it with my co-founder John Bruce on a podcast for CNBC in 2023. The term was coined in 2006 by Doc Searls, a former radio disc jockey who in 1999 had co-authored 'The Cluetrain Manifesto', an influential early document that correctly predicted the revolutionary impact the web would have on business and marketing. (The manifesto, copying Martin Luther, had ninety-five theses.) Like me, Doc had grown concerned about the online world in which we now operate – one guided not by our intentions but by what grabs our attention, like an outrageous headline or a funny meme. It's a world driven by social media, where clicks equal profit. He called this the 'attention economy'. Doc proposed replacing it with the 'intention economy', in which buyers tell the market what they want to buy, and vendors will compete to provide it.

I find Doc's framing useful. Ultimately, computers are tools. Tools can empower us, as in the early days of the web, but we must be careful not to let the inverse occur and turn the human into the tool of the computer. So imagine a world where the computer, instead of trying to distract you, actually does what you ask it to do. This would be the intention economy, of course.

Consider going on holiday. In the past you would visit a travel agent (in person) and share all your dreams for a perfect holiday. They would curate an itinerary for you and off you would go. Booking a trip online now is very different. You can find a lot of useful information regarding anywhere in the world, and the ease with which you can compare prices no doubt permits better deals. And yet there's also a cost. You might start looking for a holiday in

Portugal and you will be served more and more possible holidays in that country. But all of the information you used to share with a travel agent – your food preferences, what types of activities you like, where you have gone before – is not taken into account. A human travel agent may consider your preferences and suggest that a holiday in France might be better suited for you instead. But with personalized advertising you may not even consider that option because you have mainly been shown options for Portugal.

The attention economy is largely responsible for this deficiency. In fact, a great deal of relevant personal information that would enable richer choices is already online; it's just that the current mechanism based on attention doesn't account for all of those different data points. The paper trail of a vacation – its digital footprint – is quite significant. What does your phone know about your family holiday? The boarding passes, yes, but also the Instagram photos, the reactions to the Instagram photos, the photos that *weren't* posted to Instagram, the calendars, the star ratings of the meals, the activities, the satnav data, the exercise tracks recording your hiking, walking, running, biking and everything else. A human detective, given access to all that data, could create a pretty good picture of how satisfying the trip was, which bits worked, and which bits didn't. And an AI could do it too, but quickly.

So, say there's a new kind of travel agent. You want your next family holiday to be unforgettable, and, for the limited purpose of finding an incredible trip, you are prepared to share all that personal data from your family, just for a moment, and exclusively for the purpose of finding the best getaway, before deleting it. The goal is to obtain streamlined travel planning with personalized recommendations based on those data points. The planning could also be integrated with your digital identity for faster check-ins and security screenings. Under ultra-strict privacy controls, we can build the next amazing itinerary. This is the intention economy at work.

A focus on profit has led to social media, perhaps unintentionally, creating an attention economy. However, today's open data and open software communities offer the tools for an intention economy, and there are many of us who are trying to build it. One such person is my colleague Jesse Wright, an Australian computer science researcher and software engineer who recently took over governance of the Solid project at the Open Data Institute in London. Jesse is an idealistic and passionate programmer who's looking to engineer the future as it ought to be, to empower everyone, not merely enrich Silicon Valley venture capitalists.

I recently visited Jesse at ODI's fourth-floor offices in King's Cross, in London. We talked for a while about his progress with the Solid data protocol and our shared vision for the future. The conference room where we met offers a splendid view of the Regent's Canal, a former industrial waterway that in recent years has been redeveloped into a charming leisure area. At one point, I walked over to the window to watch the narrowboats with rooftop seating going up and down the canal, offering guided tours. Alongside them, on the canal banks, pedestrians and cyclists travelled the former towpaths where mules attached to ropes used to haul barges loaded with coal.

This evolution of infrastructure was the parallel theme to my life as a technologist. The printing press; the glowing electronics at CERN; those great steam engines I had watched as a boy . . . I had seen them come and go. As each technology evolved, the kaleidoscope turned, rearranging the power dynamics of society, leaving old actors displaced, and bringing new ones to the fore. From my work with the early web, I knew that design choices made at the start would shape the nature and values of what was to come – and once those first design choices were made, they were very hard to change.

We are at the precipice edge of the greatest technological shift of my lifetime, I thought, as I looked at the boats in the canal – the web has been transformative, the AI-powered web will be more transformative still. Yet the narrative on AI and the future more generally focuses too much on those 'doomers' and 'boomers' instead of being framed by a conversation on how we want our world to be. This time, getting the design right at the outset is more important than ever before.

Our society is now discussing how to define the terms that will govern artificial intelligence and other new technologies. Will we have Silicon Valley dictate the terms of this arrangement to us? Or might we, as citizens, take a more active role in building something empowering – through our own enterprise? The window for action here is small, and it is for this reason that Jesse, John and the rest of us in the Solid community turn to our work with a great sense of urgency.

This stuff matters. The difference between the enjoyable open web and the more predatory social media aspects is largely a design issue, a combination of the technology innovations we can create and the policy environment that enables and constrains them. I am aware of how geeky discussions of data protocols like Solid sound, but they are hugely important to my vision of a better, more humanistic world. I want people to feel optimistic about the internet – I want them to be delighted to use it! In the early days of the web, delight and surprise were everywhere, but today online life is as likely to induce anxiety as joy. By steering people away from algorithmic addiction, I hope we can reclaim that delight.

•

Artificial intelligence, of course, will be a large part of building a better world. I'm excited about AI, no question. I've been anticipating its arrival for decades. I want the use of AI to feel natural and seamless, for it to be a helper for humans all over the world. But I also see the risk that it turns into something darker, more coercive and exploitative, as happened with social media. We've got to act today to prevent that from happening. As with so much of what we need, this begins with a simple concept. It begins with trust.

The attention economy has had ruinous effects on the fabric of social trust. Since 2000, the public-relations executive Richard Edelman has tracked the levels of trust in our society through a polling initiative called the Edelman Trust Barometer. Across Europe, the US and the Middle East, the barometer measures the average person's trust in various institutions, including the media, government, business and the scientific establishment. Over the past twenty-five years, across every category, trust has collapsed – legacy media in particular has completely cratered. In its place, Edelman observes, a culture of grievance has taken root, animating all sides of political debates. Shockingly, a majority of young people globally now appears to approve of some form of what Edelman terms 'hostile activism' – harassing people online, intentionally spreading disinformation and threatening or even committing violence.*

It would probably be a mistake to attribute all of this to social media. Still, social media amplifies the speed and nature of the decline. The attention economy, which has unseated legacy media, thrives on grievance narratives, including many absurd ones. This makes building a new online economy centred on intention all the more urgent.

Fortunately – and I'm very optimistic about this – the paradigm shift of AI offers us a unique opportunity to hit the reset button. As the standards around this technology take shape, we have the opportunity not to repeat the mistakes of the social media era. You might ask, what has to happen for us to trust one another once again? But I like to turn the question on its head, and ask whom do you currently trust, and why? Why do we trust a bank with our financial information, a doctor with our medical information, or a lawyer with the sensitive details of our contracts and negotiations? In each case, there are strong cultural norms – and strong legal protections – guarding us. Doctors and lawyers take courses in professional ethics, and those who leak this information will lose their licences to practise. Banks face stiff penalties for data breaches and are aware of how much damage even a single leak can cause to their long-term business.

Now consider your search engine provider, who knows far more

* Edelman's 2025 report is here: https://www.edelman.com/sites/g/files/aatuss191/files/2025-01/2025%20Edelman%20Trust%20Barometer_U.S.%20Report.pdf

about you than a lawyer would. Instead of negotiating cultural norms to protect this data, we have instead handed it off to a tech monolith to earn profit. Imagine if your lawyer auctioned off your legal documents to the highest bidder! Online, we do this every day.

Of course, the expectation of attorney–client privilege did not arrive overnight, and it is not shared across every culture. It was the product of centuries of negotiation for the special, sovereign rights of the individual. That's why being awarded the Seoul Peace Prize for data sovereignty was significant to me, because it recognized the importance of this concept. That's my goal now, that's my passion. We have to feel like citizens, with rights, online. The battle for rights consumed the attention of earlier generations – rights for suffrage, rights for women, rights for ethnic minorities, human rights. To me, the current frontier is the battle for our rights online.

This is what the Solid community, along with others like us, are working to deliver. With Solid, we want you to trust your device, really trust it, as you might trust a lawyer, or a doctor, or a lockable safe inside your house. We rely in our typical daily life on lots of apps which we can in fact trust. Companies such as Microsoft and Apple make apps for you to use, and they try to make those apps as powerful as possible. Apple has made a point of pushing back on government requests for their users' data. Apple's business model is selling devices, not data. Similarly, if you collaborate with people by sharing a Dropbox account, then Dropbox makes its money from premium subscribers who subsidize the free accounts – not from selling user data. When people use open-source apps like Libre-Office and OpenStreetMap, they know that there is a community of people reviewing the source code to ensure it is trustworthy.

Someone recently said that they would never trust a company with their data. But would you trust your bank with your data? I trust it with my money, so I have to trust it with my data. In the same way that there is a lot of financial regulation for banks, there are privacy regulations that apply to companies that store any of your data. But is that enough? Lots of AI chatbots say they have your best interests at heart or that they won't sell the insights they get from your interactions to third parties. Why should we trust that to be true? At present there is no regulation to ensure these AI chatbots comply with their statements. We can imagine that the AIs that respect your data will become popular through their reputation and flourish as a result. However, our experience with social media is the exact opposite. Those systems that do not act in users' best interests are often the ones that have flourished.

By implementing new standards now – right now, at the start of the AI era – we can course-correct away from the exploitative use patterns that have defined the past decade. This is our chance for a 'do over', if you will. By connecting AI to a data store you trust, like Solid, or other protocols, open-source and easy to inspect, we can protect our information and dislodge the exploitative systems of communication and media that have taken hold of us over recent years.

In the next few years, AI will transform the web. In fact, it's possible it might repurpose the information pathways of the web entirely, just as new activities repurposed the waterway at Regent's Canal. As this happens, establishing data sovereignty becomes even more important. A sufficiently powerful AI might exploit you in subtle ways you can't even detect. Imagine if the same incentives for engagement that have made social media so toxic are allowed to dictate the development of AI. Imagine if we ask our smartest engineers to develop AIs which don't care about empowering users, but simply maximize our time-on-device!

What I see, meeting people in my travels and conferences, and in my communications online, is an emerging consensus that we have to build a different kind of world. I see this in the effort to create AI systems that protect your data and put the realization of your best intentions above the creation of new distractions. I also see it in the developers and users who are abandoning – or sometimes just want to abandon – platforms like X and Facebook for something pro-human. That's the essence of the 'Fediverse' – a nascent environment that has grouped together a number of open protocols, including Solid. The growth of the decentralized microblogging network Bluesky over 2024 and 2025 suggests that users desperately want something different. Bluesky is a protocol, not a website, and the eventual idea is that you can control which instance of the site you join, and the rules for moderating you want to see. Mastodon and Matrix are similarly protocol-based, with lots of different servers run by different organizations. Mastodon is my preference because the protocol, ActivityPub, is a W3C standard. In December 2023, W3C left Twitter/X for Mastodon.

Resurgent interest in pre-existing, non-exploitative technologies like RSS – the Really Simple Syndication web feed – and podcasting is encouraging as well. Podcasts deserve special attention. They make long, involved claims for your attention and they cover the full spectrum of political engagement and human culture. They can be intellectually stimulating, they can be goofy, they can be funny, scary or sad. At its best, listening to a podcast is nothing like scrolling a social media feed – it's an experience of sustained absorption, not fickle distraction. Podcasts can also be lucrative, and some podcasters become celebrities, but many seem more driven by creative passion than money. Really, anyone with a microphone can be a podcaster, and indeed some of the most famous podcasters started in exactly this way.

And as I observe this, I also observe that there is no dominant platform for podcasting, no algorithm force-feeding you certain kinds of content. The primary distribution mechanism for podcasts remains good old-fashioned RSS, the open-source, free-to-use non-exploitative distribution technology that used to be the method for disseminating blog posts and news articles. And so the podcasting environment today still resembles the web of the 1990s – stimulating, creative, collaborative and accessible. This is what the web should be.

The evolution of the Fediverse is hard to predict – there are parts of it that we can't imagine, just as, in the days of the early web, I could never have imagined something like Google search – or Github, or a webcomic like XKCD. The system will emerge organically, and I can't predict what other wonderful things will be built on top of it. With Solid, with AI, with Bluesky, Matrix and Mastodon, with RSS and a podcasting revival, I see these little shoots breaking forth from the soil – the first 'signs of spring'. I think users are ready for something else.

CHAPTER 18

Signs of Spring

In September 2024, Rosemary and I announced we were winding the Web Foundation down. When we started the foundation, in 2009, there were almost no non-profit organizations focused on the web and the social policy side of technology. Fifteen years later there were dozens, perhaps hundreds. We had accomplished a great deal, and part of our original mission – getting more people online – was satisfied. From just 20 per cent of the world's population being online in 2009, we'd moved to 70 per cent. The majority of the population of the planet was now using the web – most of them every day! Our work addressing online gender-based violence made great strides but remained unfinished, but this crusade had been taken up by many other capable organizations, including the UN. We both agreed the biggest task facing the web in 2024 was reducing the toxicity of social media, and we felt that was beyond the scope of what the Web Foundation could achieve.

As my focus turns to Solid and Inrupt, I have also stepped away from the directorship of W3C. In truth I had been delegating away my responsibilities for years, as part of a long and deliberate phase-out, to make the consortium more democratic and less autocratic. It was a relief in 2023 when the process was finally complete, and I took on the position of director emeritus. For decades W3C had used a 'hosted' model, headquartered across four international research organizations. That same year, W3C reorganized its structure, becoming a tax-exempt, not-for-profit corporation, registered in the US state of Delaware. We altered the fee schedule a bit to account for the different sizes of our partners, asking more from large, first-world for-profit companies to participate and less from non-profits and start-ups from the developing world. In my place, a board of directors was convened, comprising seven individuals elected by a straight-line vote of W3C members, plus four more nominated by the original host institutions. I was named a permanent non-voting board member, able to advise in meetings but not given a final say. And the board, while officially the highest level of authority, does not make the technical decisions: those are made by the working groups and the technical architecture board, as always.

There were many memorable times and moments along the way, and the spirit of W3C continues despite my change of role, with the same culture of earnest attention to technical excellence, inclusive processes and a whimsical delight at the weird things which have come along the road. And the same insistence on always keeping an eye on what is best for humanity at the end of the day.

The web of 2025 is very different from the web of 1992. Well, the principles are the same, but it is just so much more powerful! W3C made those changes possible, gradually, patiently, and often anonymously building the systems the web needed to thrive.

Although few outside of technological circles had even heard of the consortium, it was the patient work of thousands of technologists at W3C who successfully oversaw the evolution of the web from a few thousand computers serving static HTML pages in 1994 to the omnipresent protocol for global computing that exists across practically every connected device in the world today.

•

As I look towards the future, I see a new web taking shape. It will be decentralized, it will realize the ideal of data sovereignty, and it will put AI to good and not ill use. Remember my thought experiment, back in Chapter 12, about an AI that works for you? Well, Charlie now exists. In late 2024, we demonstrated the power of Charlie internally at Inrupt, showing how adding a completely private pod full of simulated personal data increased the AI's knowledge of the person's world.

In the default mode for Charlie, no data was shared, so when it was asked the question 'What running shoes should I wear?' it gave a general list of things anyone should take into account. When flipped to personal mode, and with consent granted to Charlie to access fitness and financial data, it gave a very considered couple of shoe suggestions, with data-driven explanations of why those would be the individual's best choice. This was so much more valuable than the un-personalized version – and so much more important than the recommendations from a corporate AI, which would hold a flash-auction to sell the result to running-shoe companies.

But getting people to *use* Solid in such a way is obviously a different conversation. So how are we going to get there? Well, I've faced this problem before. The first users of Solid are like the first users of the web. Some will approach from the quantified-self or open-data communities, in the same way that the earliest users of the web were often hypertext enthusiasts. Others will join because they are looking to solve a specific problem, or build a specific application, like when I was geotagging my skiing photos years ago. And perhaps some will grow tired of toggling between different exploitative applications on their phone and look for something more.

In one way, though, we have the wind at our back. In the old days, when I tried to explain the web to people, I first had to explain what the internet was. I don't have to do that now – and, over time, I've learned to frame ideas in concepts people already understand. At Inrupt, we've recently realized we have to stop talking about 'pods' and start talking about 'data wallets'. Calling something a pod requires an explanation, but as soon as you call something a wallet people get it. Currently, if you want to get banks and governments on board, you call it a wallet.

These aren't just nice ideas – they are taking effect in the real world. It started with Flanders – the early adopters I've already told you about – and now in Australia, a university is deploying Solid wallets for various health uses. Across the pond in North Carolina, a small organization, Datasolids, is writing the code for people to have their health data in a Solid wallet. Even in the twenty-first century, Datasolids has found in practice that the only way the doctors or hospitals can provide data is by getting them to fax it! So Datasolids have written the code to make their computer answer the fax call and put the data on the person's data wallet. In this way we may bridge the 1980s to the modern day.

The online privacy movement is stronger than ever, and it is aligned with Solid and the Fediverse. Mark Weinstein, founder of MeWe, a non-addictive, privacy-first social network, recently wrote *Restoring Our Sanity Online*, a book in which he describes a bright future in which everyone has Solid wallets as part of their re-empowered digital lives.

On the geeky end of the spectrum, Brewster Kahle, the founder of the Internet Archive, continues to organize workshops on the decentralized web. At one of Brewster's workshops, the Solid community met the LocalFirst Community, a group of open-source developers motivated very much by the same goals as Solid. One of the things which gives me confidence about the success of these movements is that they work together and are aligned. The workshops Brewster and Wendy Hanamura organize typically have not just geeks, but artists, philosophers, historians and musicians attending.

I draw inspiration from the people behind the technology, whose voices, energy and determination are driving us to a better future. Noel De Martin is a software developer who has worked developing Solid applications for the past five years. 'I still think Solid is the best thing that could happen to the Web (and maybe software in general),' he recently wrote. 'Even if there's only a slight chance that it works out, I'd regret not trying.'

The dedication of some open-source volunteers is phenomenal. A team including Alain Bourgeois in rural France and Timea Turdean in Vienna have not only kept the Solid infrastructure improving and up to date, they have also kept the systems running; they are effectively on call in case anything goes wrong for any one of hundreds of experimental users. I'm reminded of how I felt when I invented the web – they are doing this because they believe it's the right thing to do: for the vision, not for personal reward.

So change is happening from the bottom up – we've always seen that, but another sign of spring is that there is now real change happening from the top down too. It's when you have this two-way traffic that the magic happens. As we saw when drafting the Contract for the Web, spurring continuous discussions between technologists, governments and individuals is the way to create a legal framework for new technologies (like Solid) to thrive. One positive development – at least for now – is a 2024 US government regulation that says your data must be available in a 'reasonable' form. This from the same US government that not long before was mailing people their data in a PDF file written to a DVD. Requiring data to have a common language in one country might encourage other governments to format their data in a reasonable way. In the UK, the Open Data Institute has advocated for the Data Use and Access Bill which was under discussion in Parliament as of early 2025.

Social media plays a special role in the modern era: these are the public streams of consciousness through which a great many important issues are mediated. With the rise of the Fediverse, I hope that the pollution that has affected the discourse in the manipulative algorithm era is finally starting to be addressed. Instead of doom-scrolling through things that make them angry, people can engage with ideas they can contribute to. And as the discourse on a given site becomes less polarized, more people will feel at ease to express their thoughts.

These signs of spring, like seeds starting to sprout, are not necessarily coordinated. That's how the web grew too: not from a central push but through piecemeal experimental adoption. But with each new adoptee, the ecosystem gets a little stronger – and that encourages different users to connect and join forces.

I anticipate the growth of Solid as being similar to the growth of the web. You start with a small ecosystem of maybe 2,000 geeks writing code, for a user base of maybe 20,000 data enthusiasts – the kind of people who just cannot wait to link the heart rate monitor on their Oura ring with their Peloton and their phone. These people then demonstrate the utility and interoperability of these systems to their friends, while the geeks develop on-ramps for such technology to make it more accessible. Soon, successive waves of increasingly less geeky users are coming on board. Companies, individuals, governments, doctors – everyone.

•

Last summer, Rosemary and I took our sailing boat out on the lake in Canada. It's nothing too glamorous; just a 14-foot Hobie catamaran that can be operated by a single person. I placed myself on the trampoline between the hulls while Rosemary manned the tiller. We kicked off from shore under a blue sky buffeted with cauliflower-shaped clouds. The breeze was decent, and as the wind caught the sail we began to slice through the water. I shifted my centre of gravity over the side of the hull, controlling the main sheet. Because the Hobie is so low to the water, the speed can be exhilarating, especially when you catch a strong gust, and as the boat skims across the lake the downwind hull leaves a nice V-shaped dent in the water.

While the physical activity and constant adjustments kept us engaged, there were also moments of pure enjoyment. The feeling of the sun on one's back, the sound of water rushing past, the wind in the sails and the sensation of gliding over the waves can create a deep glow of connection with nature. I felt a sense of youthful optimism. I'd kept fit, and my schedule was packed. We have led full lives, certainly, met a lot of truly interesting people, been to a few events – from the quirky to the amazing. But there was still so much to be done.

Other communications technologies had come and gone, but the web had only grown. It had become a kind of default layer for daily life; a sprawling network of links now connected most of the population on the planet. From a wealthy merchant in Singapore or a smallholder farmer in Malawi, I was never more than a click or two away. Even on my sailing boat, in the middle of the water on a sunny afternoon in autumn, I could use my smartphone to contact almost anyone alive. I could deliver a package to my doorstep, or a hot lunch to my office. I could buy Rosemary flowers. I could listen to any song ever recorded and watch old clips from *Fawlty Towers*. I could do my banking or trade my portfolio or donate to UNICEF. I could read the news in any language and check the weather in any country. I could book a flight, a hotel room and a car in nearly any city in the world. All I needed was a web browser and phone reception.

I tacked the bow of the vessel around and we ducked as the boom swung across the boat. The Hobie Cat is relatively stable and easy to manoeuvre, requiring a mix of skill and intuition to keep it balanced and moving smoothly. We were sailing into the wind now, with spray kicking up through the netting onto our legs.

If the web was ubiquitous, it was not in such great shape, I reflected. Too much screen time was funnelled through the hands of too few monopolistic players. The user had been reduced to a consumable product for the advertiser. Expectations of privacy, of dignity, of sovereignty, had been corroded in the quest for profit. In authoritarian societies, the web was used as a tool of social domination and control.

The positive task at hand was to build the new, better, place, where people were empowered again. The more people I could get to use this reconceived web, the more value I perceived it would generate. I wanted Inrupt to succeed as a business, of course, and I also wanted to re-establish the primacy of the individual online. I thought of it almost as a jailbreak, releasing users from private data silos and allowing them to recreate their own connections as they wished. Systems like Solid wouldn't solve everything, of course – but my sense was that parts of the internet, and especially social media, were driving people a little crazy. I worried that those people could not imagine that the web could flip to something positive, nurturing and newly exciting. The solution was to let the light in, to open up the protocols and to design the algorithms for something other than unfettered, non-stop user engagement. Solid offered one such opportunity, to switch from the attention economy to the intention economy.

When you're sailing, you're always thinking about how you're going to get home. Putting effort into windward tacks is like investing in future freedom: once you've built up enough distance to windward, you can go anywhere you like. The wind had been on my beam – the fastest point of sail – for the first ten years of the web, and I had zoomed along. But putting W3C together, and the Web Foundation, the ODI and Inrupt, had been like investing in all the people we needed, working together to make the web better. Once that investment has been built, you have your choice of where to go.

We pivoted back to sail across the wind and began to pick up speed. The day was beautiful, and the water was dotted with kayakers, pleasure boaters, fishermen and other sailors. The rays of the sun created a patch of glittering brilliance on the water to my right, and the clouds drifted slowly across the sky, their soft edges shifting and changing shape. The world to come was beyond imagination – I could hardly believe I was alive to see it. The circle that Alan Turing had opened in the early 1950s was now closing. AI had aced the Imitation Game, and the web provided the data that made it possible.

A new age of machine intelligence is arriving. The next layer of the web wouldn't be rendered as a page but as an overlay onto the physical world. AIs trained on web data can deliver continuous streams of information and could respond in moments to spoken commands. Independent AI agents could synthesize large amounts of information and act autonomously with other web systems. Web surfing could no longer be an activity in itself but a kind of seamless informational filter on top of reality. In time, it could become a sort of sixth sense.

The immense power of such systems would be as good or evil as we permitted. It was up to us to determine what we wanted the web to be. I watched the shoreline rise and fall as the boat recaptured speed. We can still build the future we want, I thought.

There's still time to build machines that serve the human, rather than the other way around. There's still time to forge connections between people and use technology to fight for our rights. There's still time to recapture our data sovereignty. There's still time to build community. There's still time for the individual, for the child, for the parent, for the elder, for the West, for the East, for the developing world, to come together to build technology to promote the dignity of our fragile species on this isolated globe. We can do it, all of us, everyone, together.

Acknowledgements

The web is for everyone, and I would like, therefore, to thank everyone. For using the web; for, sometimes, helping to defend the web and its path to a good place. Thank you for buying this book, and for reading it, if you have got this far.

People to thank start way back, with those closest to me. My parents, as great role models in different ways; my siblings; my children, Alice, Ben, Jamie, Lyssie and Indi; and my extended family and friends. I would like to thank everyone who taught me, especially Frank Grundy and 'Daffy' Purnell at Emanuel and Prof John Moffat at Oxford, for their energy and their wisdom. I'd like to thank my fellow pupils at Sheen Mount, Emanuel and Oxford for their companionship. I also thank colleagues and friends at Plessey, DG Nash Limited and Image Computer Systems, especially Kevin Rogers and John Poole.

At CERN, everyone who helped the WWW project to form and grow, especially Mike Sendall (alas, RIP) and Peggie Rimmer, who actually enabled my working on the project at all. Thanks also to Robert Cailliau, the first convert, and more recently to Fabiola Gianotti, director-general extraordinaire.

Michael Dertouzos was the person who persuaded me to come to MIT to set up W3C, and his 'cabinet' of tech superstars at the then Laboratory for Computer Science (LCS), now CSAIL, deserve very warm thanks for their acceptance, support and companionship. They include Hal Abelson, David Clark, David Gifford, Ron Rivest, Peter Solovitz, Gerry Sussman, Steve Ward and Danny Weitzner. Rod Brooks and Daniela Rus continue the support that Michael began, so thanks to them too.

Huge thanks to Amy, my assistant, and all the wonderful W3C staff – you are too numerous to list, but your support and your strong culture have been and are a very special thing for me. Without your energy and care and attention, many of the things in this book would not have happened. Keep up the good work.

I'd like to thank all the team members and board members, and funders, of the Web Foundation in its time. You demonstrated that being on the ground in Africa and Indonesia was a key to understanding what was needed, and you achieved significant change in the world. Thank you in particular to Jono Goldstein and Kaia Miller, Brian McNeil, Marty Wikstrom, Sam Tidswell-Norrish, Alberto Ibargüen, Matt Brittin and Yonca Dervişoğlu for your support and friendship.

Thanks to the Ford Foundation for funding and Jessica Yu for directing the movie *For Everyone*, which helped to explain that the open web needs defending.

I would like to thank Louise Burke and Nigel Shadbolt, and all who have worked at the Open Data Initiative (ODI) at various times. Thanks also to Gordon Brown, Jeremy Heywood and all those involved in transforming the ODI from concept to reality.

ODI now, and Jesse Wright in particular, coordinates an ecosystem around the Solid Protocol that includes the Linked Data Storage Working Group, the Solid Community Group at W3C and also a host of open source and commercial projects and initiatives. Huge thanks to all of the volunteers, whose coordinated energy makes that whole world spin. Included in that number are Angelo Veltens, Alain Bourgeois, Timea Turdean and all the others who have worked brilliantly on bits of the SolidOS.

Thanks to John Bruce, my co-founder, and all the people at Inrupt, including Kelly O'Brien, Davi Ottenheimer and Emmet Townsend, for carrying the torch for serious commercial implementations of Solid.

For the production of this book itself, I am of course hugely indebted to my co-writer, Stephen Witt, and to my chief of staff, Grace Elcock, for her endless support and unwavering attention to detail. A great tribute and thanks to my editors at Macmillan, Mike Harpley and Alexander Star, and importantly to my agent Sarah Chalfant from the Wylie agency for her belief and vision from the beginning. This has been a long project, and I want to acknowledge others who were instrumental: Hannah Rosthschild, Jessica Bullock, Dillon Mann, Joe Ogilviy, Sarah Mortell, Olivia Larkin, Emily Treviel, Richard Tyrell and Petrina Day.

The thanks and my love end as they began, with the people closest to me, and so with Rosemary Leith, my soulmate and partner in all things, for her love and energy and patience and her strength and her ideas. This story wouldn't have happened without you.

Picture Credits

All images courtesy of Tim Berners-Lee and family unless otherwise stated.

Images on pp. 62, 77, 83 and 120 © CERN

Image on p. 79 © Punch Cartoon Library/TopFoto

Image on p. 106 courtesy of the National Center for Supercomputing Applications (NCSA) and the Board of Trustees of the University of Illinois

Image on p. 112 © Yahoo

Image on p. 134: *Internet* (Collection '"Mes Premières Découvertes"') by Gallimard Jeunesse, Jean-Philippe Chabot and Donald Grant © Gallimard Jeunesse

Image on p. 144 from *The New York Times*, 12 November © 1996 *The New York Times*. All rights reserved. Used by permission and protected by the Copyright Laws of the United States. The printing, copying, redistribution, or retransmission of this Content without express written permission is prohibited

Index

A Note About the Author

Tim Berners-Lee invented the World Wide Web in 1989 at CERN in Switzerland. Since then, through his work with the World Wide Web Consortium (W3C), the Open Data Institute (ODI), the World Wide Web Foundation; through the development of the Solid protocol; and now as CTO and cofounder of Inrupt, he has been a tireless advocate for shared standards, open web access for all, and the empowerment of individuals on the web. A firm believer in the positive power of technology, he was named in *Time* magazine's list of the most important people of the twentieth century. He has been the recipient of several awards, including the Seoul Peace Prize and the Turing Prize, widely recognized as the Nobel Prize for Computing. He was knighted in 2004 and later appointed to the Order of Merit by Her Majesty Queen Elizabeth II.